The Hong Kong Economic Policy Studies Series

TECHNOLOGY AND INDUSTRY

T0154855

TECHNOLOGY
AND
INDUSTRY

Kai-Sun Kwong

Published for

The Hong Kong Centre for Economic Research

The Better Hong Kong Foundation

The Hong Kong Economic Policy Studies Forum

by

City University of Hong Kong Press

First published 1997
Printed in Hong Kong

ISBN 962-937-002-6

Published by
City University of Hong Kong Press
City University of Hong Kong
Tat Chee Avenue, Kowloon, Hong Kong

Internet: http://www.cityu.edu.hk/upress/
E-mail: upress@cityu.edu.hk

On the cover the free-style calligraphy of *ke* ("science" in Chinese)
symbolizes the dynamic industrial sector of Hong Kong.

Contents

Detailed Chapter Contents

Foreword

The key to the economic success of Hong Kong has been a business and policy environment which is simple, predictable and transparent. Experience shows that prosperity results from policies that protect private property rights, maintain open and competitive markets, and limit the role of the government.

The rapid structural change of Hong Kong's economy in recent years has generated considerable debate over the proper role of economic policy in the future. The impending restoration of sovereignty over Hong Kong from Britain to China has further complicated the debate. Anxiety persists as to whether the pre-1997 business and policy environment of Hong Kong will continue.

During this period of economic and political transition in Hong Kong, various interested parties will be re-assessing Hong Kong's existing economic policies. Inevitably, some will advocate an agenda aimed at altering the present policy making framework to reshape the future course of public policy.

For this reason, it is of paramount importance for those familiar with economic affairs to reiterate the reasons behind the success of the economic system in the past, to identify what the challenges are for the future, to analyze and understand the economy sector by sector, and to develop appropriate policy solutions to achieve continued prosperity.

In a conversation with my colleague Y. F. Luk, we came upon the idea of inviting economists from universities in Hong Kong to take up the challenge of examining systematically the economic policy issues of Hong Kong. An expanding group of economists (The Hong Kong Economic Policy Studies Forum) met several times to give form and shape to our initial ideas. The Hong Kong Economic Policy Studies Project was then launched in 1996 with some 30 economists from the universities in Hong Kong and a few from

overseas. This is the first time in Hong Kong history that a con-
certed public effort has been undertaken by academic economists in
the territory. It represents a joint expression of our collective concerns, our hopes for a better Hong Kong, and our faith in the
economic future.

The Hong Kong Centre for Economic Research is privileged to
be co-ordinating this Project. We are particularly grateful to The
Better Hong Kong Foundation whose support and assistance has
made it possible for us to conduct the present study, the results of
which are published in this monograph. We also thank the directors
and editors of the City University of Hong Kong Press and The
Commercial Press (H.K.) for their enthusiasm and dedication
which extends far beyond the call of duty. The unfailing support of
many distinguished citizens in our endeavour and their words of
encouragement are especially gratifying.

Yue-Chim Richard Wong
Director
The Hong Kong Centre
for Economic Research

Foreword by the Series Editor

One of the most prominent changes in the Hong Kong economy over the past one and a half decades is the rapid shrinking of the manufacturing sector. From 1980 to 1994, the share of manufacturing in Hong Kong's gross domestic product dropped from 23.7% to 9.3%, while the number of persons engaged in manufacturing diminished from 900,000 to 420,000. This trend is still continuing. By September 1996 the number of employees in manufacturing had further decreased to 320,000.

It is well known that the above economic transformation in Hong Kong has been caused by the rapid economic reform and the opening of economy of Mainland China. Astute entrepreneurs in Hong Kong have deftly taken advantage of the opportunity and relocated their industrial production north of the border into China. However, from Hong Kong's point of view, what impacts would the ensuing structural change have on its economy? Must Hong Kong preserve a definite percentage of manufacturing for its economic health? How much? Is it healthy to let the future Hong Kong economy rely so heavily, if not solely, on its expanding *service* sector? These are all imminent issues to be analyzed and solved.

Furthermore, as an increasing number of economies elsewhere embark on liberalization and reform, Hong Kong faces new competition in the world market for its traditional products. At the same time, its old competitors have made special efforts to upgrade their technology level and the technology contents of their products. All these are being actively facilitated by policies of their governments. Should and can Hong Kong follow suit and catch up in the high technology products?

Some industrialists in Hong Kong have repeatedly expressed the need for government to intervene and help develop high-tech industries. What has been the role of government in technology and industry in Hong Kong? In view of the change in economic structure and of the rising competition in the world market, is there any need to redefine the role of government? Wither Hong Kong's technology and industry?

This monograph addresses these important questions and provides a useful framework for discussion. The author, Professor Kai-Sun Kwong, points out that statistical evidence clearly indicates that: the major effort of Hong Kong manufacturers in the past decade has been in the transfer of technology to the Mainland, whereas their technology level in Hong Kong has actually declined.

Professor Kwong's contribution comes also in conceptual areas. He says, in discussing manufacturing it is more fruitful to distinguish the "goods sector" from the "service sector", rather than manufacturing from services. Although most of the manufacturing process is not carried out in Hong Kong nowadays, much of the goods sector activities are still there. He concludes thus, even the service sector is becoming increasingly dominant in the economy, there can still be bright future for the goods sector. The key is that the development and market niche of Hong Kong's goods sector lie in design-intensive products.

In addition to making a strong case for a design-intensive strategy, Professor Kwong also analyzes the prerequisites for the success of the design-intensive strategy in Hong Kong, and discusses in detail the appropriate government policies to foster that strategy.

The future of industry and technology is a very important economic issue currently being debated in Hong Kong. Naturally, there are many comments and proposals by the concerned public. This monograph duly presents an insightful framework and is worthy of widespread attention.

Y. F. Luk
School of Economics and Finance
The University of Hong Kong
March 1997

Preface

I am heavily indebted to a large number of people. First of all, I cannot express adequately my gratitude to Professor Charles Kao. Everything that I know about technology can be traced ultimately to him. Amidst an exceedingly busy schedule, he graciously read through an early draft of the present monograph and gave me some sharp yet very helpful comments. I wish I could incorporate more of his suggestions, if only he could slow down from his usual delivery speed, something which only fibre optics could match. I should also thank Peter T. S. Yum of the Information Engineering Department at The Chinese University of Hong Kong (CUHK); and industrialists Allan Wong, York Liao, and Patrick Wang for helpful discussion and comments on an earlier draft. I would also like to thank P. C. Ching of the Electronic Engineering Department at CUHK, W. S. Lo of Hongkong Telcom, and Andrew Bernard and Mary Hallward of MIT for their patience and helpful discussion. Richard Wong, Director of The Hong Kong Centre for Economic Research (HKCER) and Y. F. Luk, both from the School of Economics and Finance at The University of Hong Kong, and two anonymous referees have given me elaborate and constructive comments on earlier drafts. My thanks also go to L. K. Ngai for his help in the Chinese translation and to Mon Ho of HKCER for keeping me on schedule. Finally, I am most grateful to The Better Hong Kong Foundation for their support throughout the entire project. To borrow a government adage, the Foundation has shown to the best effect the meaning of "Maximum support and minimum intervention".

Kai-Sun Kwong
Department of Economics
The Chinese University of Hong Kong

List of Illustrations

Figures

Tables

Acronyms and Abbreviations

CHAPTER 1

Introduction

The Basic Issues

Ever since the liberalization of the People's Republic of China (PRC) in 1978, the entire landscape of Hong Kong manufacturing has experienced drastic changes. Hong Kong and China will be politically integrated after the transfer of sovereignty in July 1997. By then their integration in the manufacturing arena will have been underway for almost two decades. In a nutshell, the 1970s was a decade in which Hong Kong manufacturers relied on skills and relatively low local wages of local workers. It was also a decade in which manufacturing accounted for more than 30 per cent of the total income of Hong Kong. The 1980s was a decade in which manufacturing facilities were relocated en masse to mainland China. Manufacturers took advantage of China's abundant resources to vastly expand their output. New technology was not of primary concern, as Hong Kong manufacturers were busily transferring their knowledge to their China operations. The transfer of technology saw its completion in the 1990s. In the present decade, it is hard to see any difference between what goes on in factories in Hong Kong and those in China. Not only are China operations progressing in their acquisition of manufacturing and management skills, but the vast pool of technological talent in China is also increasingly called into play. The challenge of the next decade is how Hong Kong and China can find their appropriate roles and develop their potential for the greatest mutual benefit.

Nowadays, no discussion of manufacturing can avoid the host of issues surrounding technology. Since technology across a broad

1

front — computers, microelectronics, telecommunications, bio-technology, materials, software and so on — is progressing at an unprecedented speed, no company and no government can risk lagging behind. Yet, the question of how a company or a government should shape its technology policies to suit the future is still of ongoing concern. The answer to the question of what the appropriate technology policy should be is not identical for all companies and all governments. It is important that each administration recognizes the characteristics and unique qualities of its organization, identify the unique niche into which they can fit, exploit their comparative advantages and bring out profitable results. This book shows that though Hong Kong has not been perceived as a territory that is advanced in the area of technology, there are ways in which the territory can utilize technology profitably and create an engine that propels Hong Kong into the next century.

This book aims at identifying the way forward for Hong Kong against the background of a changing manufacturing landscape and progressing technology. It is not meant to be a consultancy study. Its perspective is much broader. Its objective is to construct a setting in which questions about manufacturing and technology can be properly addressed and broad strategies identified. In addition, all suggestions for government action are directed towards the future Special Administrative Region (SAR) government of Hong Kong, not towards the Chinese Government. Under the principle of "one country, two systems", policy initiatives at the government level are best confined to the SAR government.

The Role of Hong Kong

It is often argued that Hong Kong has no role to play in the area of manufacturing and that Hong Kong can grow only as a service centre, especially as a financial centre. The distinction between manufacturing and service is a highly misleading issue and we will devote a separate section on it. What has to be pointed out here is that Hong Kong has four important roles to play when it comes to manufacturing, even if Hong Kong is to become increasingly reliant on China for its supply of land, labour and engineers.

First, Hong Kong has an advantage in terms of becoming a product design centre. Modern products have a substantial design content. Good design, which incorporates modern technology and uses new materials, can add significantly to the value of a product. This value may arise from a product's aesthetic appeal, its usefulness, or its durability. Because Hong Kong has a vast supply of management experience and a long history of exposure to foreign products, its manufacturers have a feel for what the market wants and how it can be produced cheaply and in time to hit the market. This design-intensive strategy is central to the future of Hong Kong manufacturing. Much of this book is devoted to discussing the requirements of such a strategy and to how the government can help foster it.

Besides functioning as a design centre, the second role Hong Kong can play is to act as a nerve centre for global manufacturing. This is not an aspiration, but is rather a trend that has been emerging for some time. Hong Kong manufacturers have not moved only into China, but have also relocated to Malaysia, Thailand, Vietnam and Mauritius. Advances in information technology allow the manufacturer to optimize production operations by choosing the best sites around the world at which to carry out specific operations. In global manufacturing, a nerve centre with sophisticated telecommunication links has to be set up to direct the flow of information, materials and commands. Because of the advanced telecommunication infrastructure available in Hong Kong, its central geographical location and, most importantly, the existence of managerial know-how in the territory, Hong Kong has emerged as a nerve centre of global manufacturing. To see the importance of global manufacturing, consider as an illustration a global clothing manufacturer. The production cycle begins with design patterns originating from showrooms and sales offices in the United States and Europe. These patterns are transmitted to Hong Kong for instantaneous price quotations calculated by advanced computer software. When a customer places an order, the details will be transmitted to Hong Kong and a series of operations will begin.

From the Hong Kong headquarters, materials from around the world will be sourced. The required cloth will be ordered from a textile manufacturer. The finished cloth will be shipped to another country for machining. The completed product will then be exported to the overseas market. In the process, the Hong Kong headquarters have to monitor and co-ordinate the flow of information and materials, a task that requires expert management and efficient telecommunication support. Although very little "manufacturing" takes place in Hong Kong, by acting as a nerve centre, the role of Hong Kong in global co-ordination is vital.

The third role that Hong Kong can play is to provide financing for manufacturers. In Asia, with the exception of Japan, Hong Kong has the best developed stock market, the highest concentration of banks and an abundant supply of investment capital. Whether manufacturing plants are in China or anywhere in the world, so long as there is a management presence in Hong Kong, local financial markets can be tapped. This type of cross-border financing of manufacturing activities is just beginning to emerge. In the future, banks, venture capitalists, the stock market and the government may all have a role to play in financing manufacturing companies at different stages of development and in different countries. Hong Kong's important financing role will be discussed at length in the following chapters.

Besides acting as a design centre, a nerve centre and a source of financing, Hong Kong also plays the fourth role by providing a reasonably large test market for new products. Hong Kong is at the forefront when it comes to the consumption of fashion items such as clothing, watches and clocks, sunglasses and spectacle frames, as well as in the areas of household items and electronic devices. If a product sells well in Hong Kong, it will probably sell well in the PRC and Taiwan too. This brings out a natural advantage for innovating firms. If a new product can survive a market test in Hong Kong, then the PRC, Taiwan and overseas markets can be explored with assurance. The size of the local market, which is twice that of Singapore, can also help generate the working capital necessary for expansion.

It is therefore no exaggeration to say that Hong Kong still possesses advantages for manufacturers. In light of the fact that Hong Kong is becoming more and more closely integrated with China, the territory has new roles to play. Of course, not all these roles can be classified as manufacturing roles. Indeed, acting as a financing source and as a test market is clearly not directly related to manufacturing. As for functioning as a design and nerve centre, it is argued here that these are in fact manufacturing tasks, though some people insist that they are services. To avoid unnecessary confusion over definitions, the useful distinction should be one that is made between goods and services rather than one that is made between manufacturing and services. The following section explains why this is so.

Goods Sector versus Service Sector

Conventionally speaking, economic activities can be divided into two types, those in the manufacturing sector and those in the service sector. In this study, much confusion will be avoided if economic activities in the goods sector and those in the service sector remain distinct from one another.

Firms in the goods sector engage in the various steps of goods production. These steps are in various phases which will ultimately result in the production of a saleable physical commodity. Since manufacturing is but one phase in the production process, the goods sector includes more firms than those which are conventionally classified as manufacturing firms.

Those companies in the service sector engage in the production of a service that is consumed by the final consumer the instant it is produced. Examples of such firms are retail establishments, banks, restaurants and hotels.

Accordingly, clothing design is an activity in the goods sector, although it is a service in nature. Clothing design is a production step for goods and so it is a goods activity. In contrast, the retailing of a piece of clothing into the hands of the consumer is an activity in the service sector. Retailing in the present classification is a service activity. This conceptual distinction can be generalized to other cases. For example, the packaging of products is a goods activity, whereas the transport of the packaged products from the factory to the sales point is a service activity.

The reason for adopting a goods – service classification rather than the usual manufacturing – service classification is that, in Hong Kong goods activities are much more conducive to rapid growth than are service activities. The goods sector has strategic importance. Goods can be exported on a massive scale, but services cannot. As Hong Kong is but a small economy relative to the size of the world market, exporting provides strong opportunities for economic growth. In the case of service activities, though tourism does represent the export of services, there are physical limitations to its expansion. This restriction on exports would place limits on the growth of Hong Kong as a service economy.

While most services cannot be exported, there are certain obvious exceptions. At least in theory, services such as financing and insurance can be exported. The point is that even if the demand for such services increases, these sectors need not hire much extra manpower to handle business growth. If the extra demand comes from outside Hong Kong, then such exports can lead to productivity growth. This is why some people believe that becoming a financial centre and serving the financing needs of the region is a viable alternative for Hong Kong's long term growth. While this is plausible in theory, so far its empirical support has not been easy to come by. If the financial sector is conducive to strong growth, then the labour productivity (value added divided by number of workers) of this sector should have shown strong increases in past years. Yet, available data show that labour productivity in the financial, insurance, real estate and business service sector did not show stronger increases than did the non-exporting service sectors — wholesale, retail, restaurant, transport, storage and so on. In Short, the financial sector's potential ability to bring about strong economic growth is still doubtful.

The importance of the goods sector stems from the fact that goods activities such as clothing design could grow through exporting, even though garments are not manufactured in Hong Kong. The manufacturing of clothing is a standardized process. On a global scale, clothing is manufactured where production cost is the lowest. The value that clothing output commands in the market is largely captured by designers. As clothing output increases, the value captured by designers would increase as well, in fact at the same rate. It should therefore be clear that even though designing is not usually considered a manufacturing activity, it has high growth potential. By classifying clothing design as an activity in the goods sector, it is less likely that its growth potential would be overlooked.

Design-Intensive Strategy

A central theme of this book centres around the idea that Hong Kong manufacturers should profitably follow a design-intensive strategy by focusing on finding their respective market niches in the world economy. To understand the rationale of the strategy, one should take a closer look at the stages of a production cycle.

Today, the production cycle of most goods can be divided into three phases — design, manufacturing and packaging. These phases are illustrated in Figure 1.1.[1] According to the present classification, design, manufacturing and packaging are all goods sector activities.

Depending on the particular good in question, design, manufacturing and packaging could differ in relative importance. There are goods which are very design-intensive. For example, the value of computer software is almost entirely dependent on its design. The cost of the material that goes into software production is negligible, so the economic value added of software production is exceedingly high — practically equal to the software's market value. This value added is almost entirely attributable to the efforts that have gone into the design. This example may be extreme, but the same analysis applies in differing degrees to other design-intensive products such as microprocessors, fashion, clothing, toys, electronic dictionaries and pagers.

For several reasons, design-intensive products constitute a mar-

Figure 1.1
Production Phases of a Good

Figure 1.2
Elements of Design-Intensive Strategy

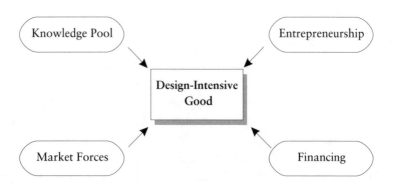

ket niche upon which Hong Kong should focus. Indeed, a number of local companies have already achieved great success in the designing of liquid crystal displays, electronic dictionaries and Chinese language pagers. First of all, Hong Kong manufacturers have a rich background in gauging the preferences of the market. The ability to judge the aesthetic appeal of a product and its usefulness to the consumer is key to design success. Second, design activities do not require a lot of space, an extremely high cost factor in the Hong Kong context. They may be knowledge intensive but they are usually not land or capital intensive. Third, because it is a creative process that adds a great deal to the value of the end product, designing can capture high returns. As wage rates in Hong Kong are significantly higher than are those in many developing economies, only high value added production can be supported locally. Fourth, looking further into the future, Hong Kong manufacturers have to look for cheaper manufacturing sites to stay competitive. These sites may be found overseas or farther inland in China. By promoting design activities in house, manufacturers develop core competencies that can be applied to future manufacturing sites.

Prerequisites of a Design-Intensive Strategy

To pursue a design-intensive strategy, several factors have to be in place in the economy. The necessary framework is illustrated in Figure 1.2. First, there must be an adequate knowledge base for the relevant technology areas. Although Hong Kong is generally perceived to be weak in this regard, the deficiency is recoverable. There are a number of remedies. Hong Kong actually has a sizable pool of locally trained engineers in various critical technology areas. What they need is practical experience in manufacturing. The best way to fill the gap is by attracting foreign direct investment. The amount of foreign technology investment in Hong Kong has lagged far behind that of some other Asian countries. To achieve this end, the government may have to take more aggressive steps in the future than it has up to now. Looking north, China provides a huge pool of technological talent, particularly in the areas of software engineering, microelectronics, biotechnology, aerospace and nuclear technology. There are two ways in which this potential can be tapped. Manufacturers can hire China engineers to work in research centres set up in China; or, they can import the engineers into Hong Kong if the Hong Kong government relaxes its immigration policy.

The second condition necessary for the successful implementation of a design-intensive strategy is that technology development be driven by decentralized initiatives from a large number of entrepreneurs. Each entrepreneur may look at things a little differently and develop slightly different products, even if the core technology is the same. Entrepreneurs are the creative people who are willing to test their ideas and bear the risk of failure. Running small start-up firms which are free from the restrictions and bureaucracies of large organizations, these entrepreneurs provide the vitality that drives every technology oriented economy.

The third condition necessary for the successful implementation of such a strategy is that products be allowed to evolve in response to the market forces. Competition in the technology market is intense. It is the consumers who decide which product is best. Consumers also indirectly exert pressure to speed up the product cycle, so that newer and better products of the existing generation of technology hit the market in a shorter amount of time. In today's market, nobody can pick winners early on. The government in particular should not attempt to do so. Such an attempt could conceivably destroy the initiatives of potential start-ups.

Finally, the financing mechanism plays an important supporting role in the area of technology development. Firms need financing at all stages. Relying on internal financing is inadequate. There should be low-friction ways to channel the economy's savings into the firms that require financing. In Hong Kong, sources of financing for young technology firms are limited. Commercial banks, the major funding source, do not usually lend to firms without adequate collateral.

Strategic Recommendations

As was pointed out in the preceding section, Hong Kong's existing capabilities do compare well to the attributes necessary for the successful implementation of a design-intensive strategy. There are four major areas in which the government can take steps to foster the development of technology and the goods sector. These will be mentioned briefly here in the following paragraphs. The details will be discussed in Chapter 6.

The first major policy area to pay attention to is the building up of technological knowledge within the government. A substantial strengthening of technological expertise within the Industry

Department would help in the implementation of all government policies related to technology and industry. To guide the efforts of the entrepreneurs and technologists, the government may consider forming a high level technology advisory committee whose objectives would be to review the state of technology and advise on those strategic areas of technology deemed to be most needed in Hong Kong. The committee might also be charged to suggest ways to develop the infrastructure of those priority areas.

Second, much more foreign direct investment should be attracted to Hong Kong in order to upgrade the practical experience of local engineers. The local engineer can learn, through working for foreign technology companies, how technology companies are managed, what products are being developed, and how design and manufacturing can be integrated. In order to attract such foreign investment, the needs of foreign companies have to be addressed and met, such as the availability of manpower, set-up costs and living environment.

Third, for the longer term, the public should be educated about technology. The government, in co-operation with private firms, should work to improve the public's awareness of technology. Exhibitions, technology fairs and creative demonstration projects may inspire the public and foster its support of new technology products. Such exhibitions will plant the seeds of future creativity among the younger generation and facilitate technology diffusion.

Finally, and for the longer term, ways to facilitate the financing of technology firms should be pursued. The government may consider establishing a venture capital company which operates along commercial principles and hires professionals in the venture capital industry as fund managers. Apart from providing support for technology industries, this initiative may help create in Hong Kong a technology related venture capitalist industry to nurture emergent technology firms. It is worth considering to create a shares market catering to small to medium sized technology firms. Such a market would allow small investors in the public to participate in the financing of technology firms with a promising growth potential.

Plan of the Book

Every country in the world is mapping out strategies in the midst of rapid technological changes. What should be produced and what should be imported? How should research be financed and subsidized? How can the technological level of manufacturing be increased? How can the necessary personnel be trained or acquired? These are just some of the issues with which every government has to contend. Hong Kong's government is no exception. The answers to these questions have to be cast in the unique context of Hong Kong. What has worked for another economy may not work for Hong Kong. While the technological challenges are common across all countries, the solutions have to be indigenous.

Consisting of seven chapters, this book attempts to provide answers to these and other related questions. Chapter 2 describes the characteristics of Hong Kong manufacturing. Hong Kong has traditionally relied on overseas orders for original equipment manufacturing. The brand name was owned by the overseas buyer and the design was, for the most part, developed overseas. It is only in recent years that more design work has been undertaken locally. Four industries stand out as the major building blocks of the manufacturing sector. They are the clothing and textile industry, the electronics industry, the plastics industry, and the metals industry. These industries, which were originally closely linked, evolved naturally and were not supported by government policies. Historically, the government insisted on supporting only infrastructural services that benefited all industries, rather than supporting selective industries. One can say that the government has never had an industrial policy in a strict sense. As a result, manufacturers generally had to rely on themselves. Financing was no exception; manufacturers were predominantly privately financed. Only those few that had collateral received support from banks. These features of the manufacturing sector are discussed in detail in Chapter 2 and they form the foundation upon which the discussion in Chapter 4 on Hong Kong's future technological and industrial strategies is based.

It was pointed out earlier in this introductory chapter that the 1980s was a decade in which Hong Kong manufacturers rapidly transferred their knowledge to China. The utilization of new technology inside China was not their primary concern. In some cases, the production methods adopted in China were more labour intensive than those that had been used in Hong Kong. There is evidence that the technology level of Hong Kong manufacturers in the 1980s was actually in decline. Chapter 3 explains how this came about. In fact, while fewer workers were employed in the factories of Hong Kong (not counting those physically employed in China), the quantity of both machinery and imported unfinished goods used in Hong Kong increased significantly. It was a case of using a lot of resources to produce roughly the same amount of goods. For this reason, the measured level of technology declined between 1984 and 1993. The implication of this trend is that Hong Kong manufacturers, who in most cases have production plants in China, urgently need to upgrade technology and engage in activities of higher value added. This is not a new call for action. There have been calls for higher value added production since the 1970s. What is different now is a whole new set of external factors: the rapid diffusion of advance technology worldwide, the globalization of manufacturing, the liberalization of many low wage areas, and Hong Kong's integration with China. These changes necessitate a new assessment of Hong Kong's role and the niche upon which the territory should focus.

Chapter 4, which is the core chapter of this book, argues why Hong Kong manufacturers should focus on a design-intensive niche. While technology is progressing rapidly, the adaptation of technology to serve consumers better is often the most challenging aspect of technology development. Bridging the gap between technology and consumers is not only an ongoing challenge but also a lucrative opportunity. In many cases, the gap can be filled by better design. Chapter 4 sets out in detail the reasons for which this niche is appropriate for Hong Kong and explains in detail the prerequisites for success in this area.

Chapter 5 is a case study showing the way in which the design-intensive strategy has been carried out successfully in Hong Kong.

The chapter draws upon the experience of several Hong Kong companies. Their stories show that utilizing technology to pursue a design-intensive strategy is not a far-fetched dream. To a number of companies, it is already a reality.

Up to this point, little has been said about the proper role of the government in the new era. A discussion of the policies that the government should pursue is deferred to Chapter 6, after the strategic issues have been made clear. Chapter 6 discusses in detail the major policy areas — the building up of technological expertise within the government, the inducement of foreign direct investment, the enhancement of the public's awareness of technology, and the creation of a better financing mechanism. Chapter 6 also addresses, within the present framework, some often debated issues concerning industry and technology policy.

Before moving on, one final remark is in order. In some sense, the transition problem Hong Kong now faces is not unique. Other major cities of the world — New York, London and Paris — have transformed themselves from manufacturing bases into knowledge intensive, service oriented centres. The fact that these cities have survived their transitions suggests that Hong Kong need not worry unduly.[2] One should not overlook, however, the fact that Hong Kong people have much less mobility than do the inhabitants of many other cities. A Londoner stays British even if he moves out of London, for he continues to speak the same language, pay roughly the same taxes and possess the same citizen rights. Workers displaced from declining industries can move to other cities in their own country. Housing prices, labour wages and other input costs would adjust in accordance with changes in the market. The transition is therefore relatively smooth. Hong Kong, however, is less fortunate in that its citizens' mobility is much more restricted. In any case, in the years to come, Hong Kong has to keep on growing continuously in order to keep workers employed without sacrificing living standards. The depth and seriousness of the 1997 transition problem render Hong Kong's situation unique. While not being the core issue of this book, the transitional issues are variously addressed in the pages that follow.

Notes

1. The reader may find that breaking down the production process into separate phases is similar to the idea of "value chain" in Porter (1990). Porter (1990) also stresses the importance of finding a high value added niche along the value chain, but his argument is not focused on technology development.

2. I thank an anonymous referee for pointing this out.

CHAPTER 2

Characteristics of Hong Kong Manufacturing

This chapter describes the main characteristics of the manufacturing sector in Hong Kong. Familiarity with these characteristics is essential for the formulation of appropriate industry and technology policies. It was pointed out in Chapter 1 that our discussion is going to focus on the goods sector rather than on the sector that engages in manufacturing activities only. Since it has been only in recent years that Hong Kong manufacturers have engaged in activities other than pure manufacturing, treating the manufacturing sector and the goods sector as equivalent should not create much confusion in this chapter.

In Hong Kong, the evolution of the manufacturing sector has largely been shaped by market forces. There are four major industries in Hong Kong — the clothing and textile industry, the electronics industry, the plastics industry and the metals industry. That these industries have achieved world dominance has been a natural outcome of a combination of human, physical and economic factors unique to Hong Kong. The Hong Kong government's role in the evolution of this dominant position has been limited. The government has all along seen its role as a provider of infrastructural support for all industries. It has never attempted to pick winners and boost the development of specific industries.

Hong Kong manufacturing has been particularly strong in a number of areas. Its major competitive advantages are flexibility, material sourcing capabilities, short delivery time, deep knowledge of the market and expertise in production management. These qualities have helped Hong Kong manufacturers secure orders which are small in volume and high in quality, and which require quick delivery to the marketplace. Hong Kong's capability in these respects fits in very well with the just-in-time inventory management practice which most overseas buyers adopted in the early 1980s.

Another major factor that has obviously helped Hong Kong manufacturers is the liberalization of China. With the exception of certain goods that were restricted under export quotas, China's abundant labour and land resources have allowed these manufacturers to vastly expand their outputs. An owner of a factory employing 100 in Hong Kong might suddenly become the de facto owner of a factory employing 5000 in China. As a result, Hong Kong manufacturers have become competitive with other developing countries in the region. In an era of rapid globalization in manufacturing, the resources that China offers have allowed manufacturers to reap profits without much exploration overseas.

Within Hong Kong, manufacturing firms have always been small and they have become even smaller in recent years. For instance, government data show that in the early 1980s the average size of a Hong Kong firm was rather constant, standing at around 18 workers per firm. By 1990 the figure had dropped to 14.8 workers per firm and in 1994 the corresponding figure was 12.5. The relocation of labour intensive production processes to China has been the major reason behind this downsizing of manufacturing firms.

Manufacturing firms in Hong Kong are typically owner managed. This fact is consistent with the entrepreneurial nature of the Hong Kong economy. The owners are usually capable managers with an intimate knowledge of every facet of their business. Workers are generally skillful and the mode of production is skill intensive. Research and development in the formal sense are, however, not the strength of owners or workers. They basically carry out production operations as required by buyers. Even when manufacturers set up larger factories in China, basic management methods remained the same. In the early years, managers were sent from Hong Kong to start up China operations. Because of the proximity of southern China to Hong Kong, any unusual problems that arose could quickly be resolved by the Hong Kong owners in person.

In Hong Kong, financing of business expansion comes mainly from internal and private sources. The banking industry has not been very supportive of manufacturing businesses. Those banks which do lend to manufacturers are almost as a rule protected by cash deposits and real estate collateral. This problem is not unique to Hong Kong. Small businesses in countries around the world often find bank loans difficult to secure.

In terms of the source of technology, manufacturers have largely acquired their technology from overseas investors. Foreign direct investment is probably the single most important source of technology. As foreign investors have a stake in production facilities, they only export technology that is appropriate in the Hong Kong context. While Hong Kong manufacturers do not commonly conduct technology research, they are keen to acquire modern machinery. In some cases, this has been a good way of utilizing technology to improve production efficiency and quality. There have also been cases, however, in which inappropriate machinery was acquired or in which the machinery acquired was so advanced that it did not operate as well as expected.

Within the borders of Hong Kong, manufacturing is a shrinking sector. According to government statistics, the number of workers employed locally in this sector has been diminishing, as has the share of manufacturing in the gross domestic product. The opening up of China has given manufacturers access to low cost production facilities in southern China. To most manufacturers, the China Hong Kong boundary is non-existent. Today, China is no longer a manufacturing site that offers cheap production alone. It has managers, technicians and engineers whose capabilities are comparable to their Hong Kong counterparts. Some manufacturers even find it difficult to come up with a reason for keeping a base in Hong Kong. It should be emphasized here, however, that though it is not widely recognized, Hong Kong does have comparative advantages in goods sector activities, especially in acting as a global nerve centre and as a design centre. The latter role in particular dovetails into the design-intensive strategy. It is by far the more important of the two roles. It is also the role that will be emphasized and discussed in the following chapters.

The Four Major Industries

During 40 years of industrialization, the development of Hong Kong's manufacturing sector followed basically a natural path of evolution which was shaped by the world market and the territory's economic advantages. Because of the high price of land and investors' short planning horizon, heavy industries could not develop. Indeed, most manufacturing plants in Hong Kong are small establishments accommodated in multi-storeyed factory buildings. The emphasis on light industries carried over to the era of massive relocation to China about 15 years ago. The profile of industry types in southern China today is therefore very close to that in Hong Kong.

Four industries stand out as the building blocks of Hong Kong manufacturing (Figure 2.1). They are the clothing and textile industry, the electronics industry, the plastics industry and the metals industry. Within each industry the range of products is quite wide.[1] Products in the clothing and textile industry include cotton, silk and synthetic fabrics, surface treatment and dyeing of fabrics, all types of outer and inner garments, knitted garments, leather goods and footwear. Products in the electronics industry include audio and video equipment such as tape recorders and TV sets; scientific equipment such as calculators; electronic components such as printed circuit boards and integrated circuits; telecommunications equipment such as telephones and facsimile machines; and a host of small electronic consumer products such as smoke detectors, weighing scales and electronic diaries. Products in the plastics industry include a wide range of casings and parts for electronic and electrical equipment, watches and clocks, plastic household goods, soft toys and miscellaneous items such as plastic bags and bottles. Products in the metals industry include parts and casings for electri-

Figure 2.1
Distribution of Value-added in Major Manufacturing Industries

1984

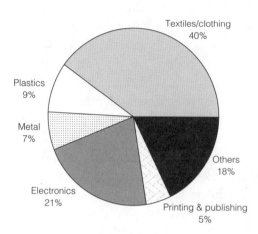

1993

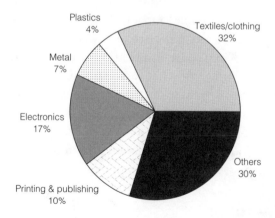

Source: *Hong Kong Annual Digest of Statistics*, 1986, 1995

cal and electronic equipment, machine tools, moulds, metal toys, watch bands, surface treatment such as electroplating and a wide range of household items. Besides these four major industries, two upcoming industries that have been growing quickly are the printing and publishing industry and the jewellery industry. One probable factor underlying their increased importance is these two industries' heavy reliance on design capabilities.

The close linkages between the four major industries are apparent. The close relationship between the clothing and textile industries is the easiest to understand. Such a close supplier – user relationship exists among the electronics, plastics and metals industries as well. A facsimile machine, for example, is made with products from the electronics, plastics and metals industries. Other products such as clocks, video, cameras and calculators are also examples of cross-industry products.

That Hong Kong's successful products draw primarily on the manufacturing capability of a number of existing industries is a point that needs to be emphasized, particularly in policy discussions. The design and manufacturing of a product require the expertise available in a number of industries. Participating firms must interact closely in the development of that product. They form networks about which people outside their industry know little. These networks are an integral part of the manufacturing infrastructure of Hong Kong. The implication is that people in the business are the best judges of what can be produced and what cannot. Ultimately, only people in the business, and not the government, are capable of answering the question of which new products can be profitably developed in Hong Kong. The decentralized nature of the industries makes it very hard for the government to force its way in boosting the development of a specific industry.

Mode of Operation

Despite 40 years of industrialization, the organization of a typical manufacturing firm in Hong Kong has not undergone significant changes. The typical firm today is still owner managed, entrepreneurial and skill intensive. Some firms might have a large plant in China employing hundreds or thousands of workers. But the basic structure of an owner-supervisor-worker hierarchy has remained largely unchanged. As the China operations are just north of the China – Hong Kong border, usually within a few hours travelling time over land, the owner-managers need not delegate complete authority to the managers they hire in China.

This type of organizational structure is very well suited to the sort of business in which manufacturers typically engage — original equipment manufacturing (OEM). In this kind of business, the overseas buyer has the design and specification for a new product, say a pair of jeans, a Ninja turtle toy, a pair of athletic shoes or a pair of spectacle frames. The buyer looks around the world for a place in which the product can be most economically produced. The search frequently ends in Hong Kong because of Hong Kong manufacturers' all round skills, flexibility and reliability, and because of China's low production cost. Overseas buyers usually own the brand name and the design of the product. They also own the distribution network and they do the promotion. The advantages to the Hong Kong manufacturers in the OEM business is that they bear very little market risk, and the capital requirement beyond that of the necessary machinery is small. So long as the buyer pays up, there is almost no risk involved. On the negative side, however, the profit margin is low because Hong Kong has to compete with low cost production sites elsewhere. If it were not for China, manufacturers could have been driven out of the market long ago by the high wages in Hong Kong.

The key to the OEM business is volume. Manufacturers need to maintain their plants at a very high rate of utilization in order to generate profit. To do that, they have to keep a wide network of business connections and to continuously explore new markets, both of which require good telecommunication facilities. As both the cause and the effect of OEM business, the telecommunications infrastructure in Hong Kong is one of the most highly developed in the world.

It is well known that typical manufacturers in Hong Kong

spend little on research and development (R&D). The reason for this is that in the OEM business they do not need to conduct R&D. The key to OEM is low wages and high volume. The overseas buyers take care of R&D. The local manufacturer's concern is to turn out products quickly and to meet quality standards. At a time when buyers are striving to cut inventory costs by shortening the order-delivery cycle, Hong Kong's efficiency and reliability command a premium. This quick response advantage is not lost even though a great deal of manufacturing takes place in southern China. This has been possible thanks to the proximity of this production base to Hong Kong, especially to the direct highway and rail links between the two.

Looking into the future, as manufacturers move to higher value added activities, such as the kind of design work emphasized in this book, R&D will become indispensable. R&D work has to take place in both Hong Kong and China. The division of labour may go this way. Hong Kong will have an advantage in research to solve problems related to the end-product market, such as aesthetic design and technology adoption in certain areas. This would be especially profitable if a critical mass of foreign direct investment is injected into Hong Kong. China will on the other hand have an advantage in research areas further away from the end-product markets, such as programming, software engineering, new materials and artificial intelligence. This division of labour in research will become clear as the post-1997 China – Hong Kong relationship evolves and as the two administrative areas assume their respective roles.

Financing

Manufacturers in Hong Kong are largely internally financed, either from retained earnings or from owners' personal savings. Most manufacturers do not qualify for external financing. Even heavy investments required to set up large operations in China are usually privately financed.

In the capital market, one main source of financing is bank lending. Yet, very few manufacturers receive bank loans, because they lack collaterals. Those who do qualify for a bank loan are usually the ones who own real estate properties and are willing to pledge them against the loan. For manufacturers who do not own real properties, their largest assets are probably their human capital and their machinery, neither of which offer downside protection to the bank should the manufacturer default on the loan.

One source of external financing which is available to a limited few is the stock market. The really successful manufacturer who has a continuous profit record for three or more years may receive sponsorship to go public on the Stock Exchange. Nevertheless, the original owners usually have to hold a majority stake in the company after going public. The typical arrangement is for an original owner to put 25 per cent up for public investors at the time of the initial listing and to keep the remaining 75 per cent to himself. The Hong Kong Stock Exchange requires a minimum of 25 per cent of the shares to remain in constant circulation.

It is generally understood that start-up firms have a great deal of difficulty in raising funds. The problem is particularly serious for technology start-up firms. These firms are typically strong in human capital but weak in collateral. Unlike in the United States, there is no venture capital in Hong Kong that supports technology firms. The government, until recently, has had no scheme to support them, either. Even the recent funding schemes have their limitations. This is not to say that Hong Kong lacks investment capital. Besides institutions, there are many high net worth individuals in the territory who are on the lookout for good investments. What is lacking is a mechanism that brings the two parties together. This financing issue will be picked up again in the following chapters.

The Government's Role

The government has always claimed that its policy towards industry is one of "positive non-intervention", which is a rather vague term indeed. What it means in practice is that the government supports industrial growth through infrastructure provision (roads, energy, water and sewerage) but avoids direct subsidies and other means of intervention that aim at changing the composition of the industrial portfolio of a free market.

The government's position is probably unique among industrializing economies. The governments of Singapore, Taiwan and South Korea have a much stronger hand in directing industrial development.[2] They have well defined and well publicized industrial and technology policies.

In Hong Kong, the government does not believe that its intervention can benefit the economy as a whole. Moreover, even if the government believed in intervention, the impact would still be limited. The reality is that the government is small, inexperienced in running industries and lacking in technological expertise. Intervention would either be ineffectual or disastrous. This does not mean, however, that the government should not formulate industrial and technology policies. As will be explained in Chapter 6, there are a number of initiatives that the government can implement with good results, including the strengthening of the Industry Department, the attraction of foreign direct investment, the enhancement of technological awareness in the public, and the creation of better financing mechanisms.

Technology Transfer

There are two parties to a technology transfer — the transferor and the transferee. In the case of Hong Kong, technology transferors are usually foreign direct investors. Although there are semi-government organizations that set up laboratories to facilitate transfer, their roles are relatively minor. Foreign investors choose Hong Kong as a production site for a wide variety of reasons, some of which are: relatively low wages for the same worker quality, low tax rates, free capital flow, efficient shipment facilities and proximity to the market. The transferees are usually experienced production engineers who are not necessarily highly educated, but who have the capability to assimilate the technologies and carry them out locally. In the past, production technologies such as plastic injection moulding, printed circuit board manufacturing, surface mounting, precision tool making and combustion engine design and assembly have been acquired in this manner.

The main benefit of acquiring technology through foreign investment is that it is close to market needs. Because technology transferors are also investors, they have an incentive to transfer technologies that are appropriate to the Hong Kong context. Even if the technology is not the best for Hong Kong from the social point of view, it may be optimal from the firm's point of view. Furthermore, the best way to learn about a technology is to see it in action. The experience acquired from "learning by doing" cannot be easily transferred through books and documentation. It can only be acquired through observation and practice.[3] In Hong Kong, engineers learn quickly and they reach high quality standards. According to some foreign investors, it is common for experienced local engineers to achieve output quality that surpasses the standards set out by the machinery vendors. Worker quality is a major factor that has attracted foreign investors to Hong Kong.

In recent years, Hong Kong has lost much of its appeal to foreign direct investors. The reasons for this are complex. The attractiveness of China, the proactive policies of other countries such as Singapore and Malaysia, high local costs and the lack of government initiative in Hong Kong may all play a part. This is a weak area that the government urgently needs to address.

Industry Performance

According to official statistics, the manufacturing sector within Hong Kong has been gradually shrinking.[4] In 1978, the year China began to liberalize, the manufacturing sector accounted for 33.8 per cent of gross domestic product. In 1993 the corresponding figure dropped to 10.5 per cent. The decline in share has been particularly marked since 1988, as many manufacturing firms ceased their manufacturing activities in Hong Kong and became trading firms. Most of those that continued manufacturing moved their facilities to southern China. It has been estimated that up to four million workers in southern China are now engaging in manufacturing activities that originate from Hong Kong manufacturers.

Today, even so called manufacturing firms are doing little manufacturing in Hong Kong. They are now principally engaged in logistics, material sourcing, designing, accounting, quality control, packaging and trading. One exception, however, is clothing manufacturers. Because clothing manufactured in China is subject to export quota control, a great deal of manufacturing in the clothing industry has remained in Hong Kong. The export quotas for Hong Kong made clothing are widely distributed among large manufacturers, and there is a ready market for the transfer of such quotas.

In the first few years after the opening up of China, products travelled a rather circuitous route. Many manufacturers stationed their designing, sourcing, sales and management functions in Hong Kong, while the raw materials would go to China for processing. The semi-manufactured articles would be shipped back to Hong Kong, where they would undergo quality control and packaging before finally being exported. Gradually, as operators in China accumulated production experience and moved up the learning curve, many products began to be exported directly from China, or via Hong Kong as a transshipment point, without further processing in Hong Kong.

An interesting question arises as to whether manufacturers have been upgrading their production technology. It is quite conceivable that in the years during which Hong Kong manufacturers were transferring their knowledge to China, their operations in Hong Kong and China combined suffered a decline in productivity. One can interpret this decline in productivity as a decline in technology level, since during this period more resources than before were needed to produce the same output.

The hypothesis that manufacturing technology has been declining is support in a recent study of the total factor productivity of Hong Kong manufacturers.[5] The study assumes that growth in the value of manufacturing output depends on growth in five input categories — worker hours, stock of machinery and equipment, factory space, value of intermediate inputs, and operating expenses

such as energy and transportation. Under this measurement framework, if all inputs are found to grow by x per cent, but output grows
by less than x per cent, then total factor productivity is decreased.

Total factor productivity could decrease for a number of reasons. If raw materials were wasted, then more raw materials would
have to be used to produce the same amount of output. If the workers employed were less skilled than were those in the past, then more
worker hours would be required. If the quality of output declined, it
would have to be sold at a lower price, and so output value would
drop. If the same product were produced year after year, the falling
market price of that product would reduce the value of output. If
new machinery were acquired but was not put to good use, then the
investment would be wasted. Any of these factors could lead to a
measured decline in total factor productivity. Roughly speaking,
these are indications of a decline in technology level.

The study found that in the decade since 1984, the total factor
productivity of the Hong Kong manufacturing sector exhibited a
slight decline. The reasons for this trend have yet to be confirmed.
The data show that during the period under study the value of output was steady, and worker hours decreased rapidly, but the stock
of machinery and equipment and the flow of semi-manufactured
goods (mostly from China) rapidly increased. The net result was
that the expansion of machinery and semi-manufactured goods
dominated the contraction of worker population, and the manufacturing sector was becoming less efficient. Manufacturers are now
using more inputs to produce the same amount of output. Technology level appears to be in decline.

Nevertheless, the decline in total factor productivity has no direct impact on the profitability of manufacturers, at least not in the
short run. Though they may consume excessive amounts of semi-
manufactured goods from China, their production costs may be
low enough to earn them profits. The worry is about in the longer
run effects. Eventually, manufacturers will have to upgrade
technology. They will have to find a niche in which the unique
advantages of both China and Hong Kong can be put to good use.
This issue will be addressed in the remainder of this book.

Notes

1. The Industry Department commissions consultancy studies on the four
    major industries in a four year cycle. Much of the material in the following
    is extracted from those consultancy reports.

2. South Korea (through the Korea Institute of Science and Technology),
    Taiwan (through the Industrial Technology Research Institute), Singa-
    pore (through the government established institutes for microelectronics,
    advanced materials and biotechnology), Malaysia and Thailand have well
    defined technology policies. For detailed country studies, see the collected
    volume edited by Simon (1995).

3. The idea originates from the concept of "tacit knowledge" introduced by
    Polanyi (1967) and popularized by Nelson and Winter (1982). Romer
    (1992), in his discussion of the importance of technology transfer through
    direct investment, also evokes the idea.

4. Data extracted from *Annual Digest of Statistics*, 1995 edition

5. See Kwong, Lau, and Lin (1996).

CHAPTER 3

A Quantitative Analysis of the Manufacturing Sector

Since Hong Kong's manufacturing sector has been undergoing rapid changes, it might be useful to go a bit deeper into a quantitative analysis of those changes. An analysis of the trends in output and input changes would serve as a basis for understanding why the measured technology level of Hong Kong has shown a steady decline in the past decade or so. This decline points to a clear need for the goods sector to target an appropriate niche. The data also show the cost structure of Hong Kong manufacturing in quantitative terms. These findings underlie the strategic considerations in the next chapter. The material in this chapter is largely based on Kwong, Lau, and Lin (1996).

The data used in this study were collected by the government through the annual Survey of Industrial Production. This series started in 1984 and the latest year available is 1993. The format of the survey in the years between 1984 and 1993 remained largely unchanged, with the exception that industries were reclassified in 1989. During this period, the relocation of manufacturing facilities was occurring at full speed. Government surveys covered only firms that were conducting manufacturing operations in Hong Kong. During the period, many firms ceased their manufacturing operations in Hong Kong. Although these firms might still maintain an office for storage and for trading activities, they were taken off the government sample.

The output of the manufacturing sector can be measured in a number of ways. One frequently used measure is value added.

Value added is defined as the market value of goods produced, that is, output minus the payments to input materials and semi-manufactured products used, minus rental for land, buildings and machinery and operating expenses. The value added of the manufacturing sector is an important component of the gross domestic product of Hong Kong, the other important component being that of the service sectors.

For an important reason, value added is not an appropriate way to measure the output of manufacturing in Hong Kong, especially when measuring the technology level is considered. As Hong Kong manufacturers moved their operations out of Hong Kong, particularly to China, their profits became higher than they otherwise would have been. This led to a higher value added in Hong Kong. As this occurred without any increase in inputs used in Hong Kong, it makes it appear as if manufacturers had become more productive. This perception could lead to the belief that technology in Hong Kong had improved. In reality, however, a large amount of low cost labour in China was used, and its outputs were shipped back to Hong Kong in the form of semi-manufactured goods. As these goods were produced at low costs, even though they were subtracted out in the measurement of value added, an increase in value added would still occur in the absence of increased use of inputs, leading to a measured improvement in technology. In one notable study, Young (1992) finds that using the value added concept, the Hong Kong economy exhibited significant improvement in total factor productivity. Much of this improvement could have been due to the relocation of manufacturing rather than to a genuine technology upgrade.

The alternative is to measure output directly. The problem is that the output of one industry may be the input of another, so that a summation of the output of all manufacturing firms is subject to a double counting error. It is impossible to make all the necessary adjustments to eliminate this error. There are certain obvious cases such as that of the output of the electroplating, surface treatment and dyeing industries, in which the outputs of these industries can easily be removed from the output of the manufacturing sector.

On the input side, several key inputs can be identified — labour, factory space, capital (i.e., plants, equipment and other kinds of fixed assets besides land and buildings), raw materials and intermediate inputs and operating expenses. Each input commands compensation in the form of factor payment. As capital is typically owned by the owners of the firm, payment to capital includes the accounting profit of the firm. Since the five inputs exhaust the list of inputs, the sum of the payments to the five factors equals the value of goods produced.

The measurement of technology progress requires data on the growth of the quantity of each input and the share of each input in the total value of goods produced. The details of the measurement process are a technical matter and will not be dealt with here. For our purposes, it suffices to say that the government surveys undertaken from 1984 to 1993 provide sufficient data to allow a meaningful comparison of technology levels throughout the period. One can measure not only the level of the manufacturing sector as a whole, but also the levels of each of the component industries. Generally speaking, the measurement process would show an improvement in technology if (the logarithm of) output value grows faster than the weighted average of (the logarithm of) the growth rates of individual inputs.

Among the five inputs, intermediate inputs accounted for the largest fraction of output value. In 1984 intermediate inputs accounted for roughly 35 per cent of output value, reflecting the prevailing OEM mode of business. More importantly, this percentage increased throughout the decade, resulting in 52 per cent of output value being attributable to intermediate inputs in 1993. This is an indication of the extent to which part of the manufacturing process relocated out of Hong Kong. On the other hand, as manufacturing activities moved out, one would expect the percentage attributable to capital to decrease. This was indeed the case, for the change in capital share was almost the mirror image of that of intermediate inputs. In 1984 capital accounted for roughly 31 per cent of output value. The figure dropped to roughly 17 per cent in 1993.

Despite the large changes in intermediate inputs and capital,

Figure 3.1
Total Factor Productivity (TFP) and Labour Productivity

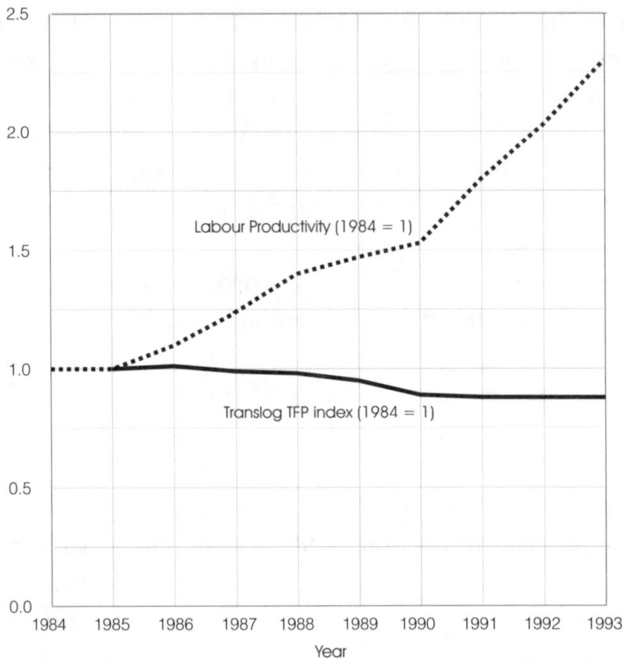

Source: Kwong, Lau, and Lin (1996).

intermediate inputs and capital combined accounted for two thirds of output value, a fraction that stayed rather constant throughout the decade.

Labour intensity was, as is shown by the data, not particularly high. Throughout the decade, the fraction of output value attributable to labour remained rather steady, standing at about 18 per cent. In practical terms, therefore, a 10 per cent increase in wage rates would have increased manufacturing costs by about 1.8 per cent — not a huge margin. It seems safe to say that the rapid relocation out of Hong Kong was a result of the liberalization of China and other countries in the region and not of changes in the local labour market.

Factory space is a valuable resource in Hong Kong. Many people have the impression that such space is an important cost component in Hong Kong manufacturing. The data offer little support for this view. Factory space accounted for around 2 to 3 per cent of output value during the decade. In practical terms, therefore, a 10 per cent increase in factory rentals would increase manufacturing costs by about 0.25 per cent. It is, therefore, somewhat misleading to say that high land prices have driven manufacturing activities away. Again, it was the liberalization of China and other countries with substantial land resources that attracted such activities and caused them to move out of Hong Kong. This observation is perhaps relevant to the ongoing debate over whether government provision of industrial land at a below market level could result in manufacturing activities remaining within Hong Kong's borders. The data suggest that such a policy by itself would likely have a negligible effect.

Based on the survey data, the state of technological progress in the manufacturing sector appears rather gloomy. What came out of the model of technology level measurement was that technology was generally in decline between 1984 and 1993. The decline was particularly rapid after 1989, as is shown in Figure 3.1. In concrete terms, if the manufacturing sector used the same resources in 1993 as it did in 1984, it could have produced only 87 per cent of its 1984 output. In other words, 13 per cent of the output was lost through a decline in technology. Manufacturers were generally not able to produce goods of higher quality and market value. Instead, they transferred their technology to low cost countries and made use of their low cost resources to produce standard goods. The use of large amounts of foreign resources implies that technology has not been improving. The data suggest that Hong Kong manufacturers are just growing larger, not better.

In Figure 3.1, the change in labour productivity during the decade is also shown. Labour productivity is defined as the output value of the manufacturing sector divided by the number of worker hours. Unlike total factor productivity (TFP), labour productivity increased rapidly during the decade. The reason for this is that the number of workers declined rapidly, but the output value did not. Of course, this notion of labour productivity is rather misleading in

Chapter 3

Figure 3.2
Changes in Output and Input Levels

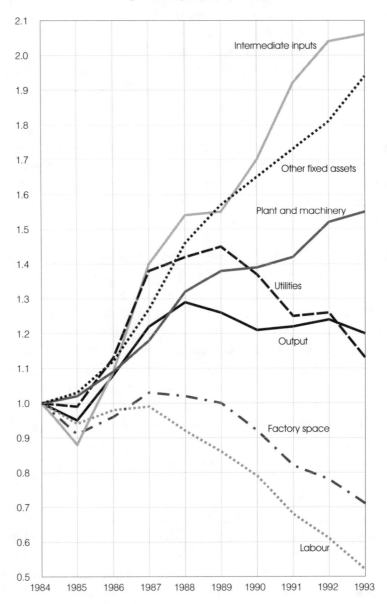

Source: Kwong, Lau, and Lin (1996)

the present context, because workers outside Hong Kong have not been taken into account.

To discover the underpinnings of the decline in technology, one has to investigate the changes in each of the inputs. Figure 3.2 can shed some light on this topic. In the figure, output and inputs throughout the decade have been deflated by appropriate price indexes, and each path has been normalized to begin at unity in 1984. For the output series, output level increased slightly from 1984 to 1988 and then flattened out. Worker hours (within Hong Kong) contracted continuously, dropping by 1993 to half the 1984 level. Also dropping was factory space, although the plunge was not as drastic as was that in worker hours. All the other inputs showed significant increases. In particular, intermediate inputs rose by more than 100 per cent, "other fixed assets" by almost 100 per cent, and plants and machinery by more than 50 per cent. It is clear that the manufacturing sector has been consuming increasingly large amount of inputs in the absence of substantial output growth. This disparity is readily captured in the model of total factor productivity measurement (Figure 3.1), and it manifests itself in a decline in measured technology level.

So far, the unit of analysis has been the manufacturing sector as a whole. One can also apply the measurement framework to each of the component industries. The findings present a picture which is quite diverse. Some industries have improved in the area of technology, some have declined, while others have been stagnant. In the first category, the notable industries are the food manufacturing industry, the beverage industry, the non-rubber footwear industry, the printing and publishing industry, the electronic components industry and the electronic toys industry. In the second category, the notable industries are the plastics industry, the metals industry and the chemicals industry. In the third category, the notable industries are the wearing apparel industry, the leather products industry and the scientific equipment industry. These findings accord well with common observations. The food, beverage, printing and publishing, electronic components and electronic toys industries have been able to grow successfully through quality upgrade and new product creation. The most important element that they have injected into their businesses is creative design. The design content of their products enhances their market value and increases total factor productivity. These are signs of growth. The data do lend support, therefore, to our argument that design-intensive goods should be an important aspect of future Hong Kong manufacturing.

CHAPTER 4

Strategic Considerations

In an era during which a diverse range of technologies are evolving at great speed, many individuals, firms, cities and even nations are looking for a niche in which they can capture the powers of technology and move ahead of others. Technology policy is taking on greater strategic importance and is encompassing industrial policy. Formulating a technology policy is a challenging task for any government, because the economist's invisible hand may not be very effective in promoting technology development. Technology, on a general level, is like education; it has a strong public good nature. One cannot rely on market prices to allocate resources to the production of technological knowledge. Furthermore, technology development requires focus, leadership and co-ordination. Based on the lessons learned by many countries, successful development in technology seems to require government leadership and planning.[1] But even if the government plays a leadership role, technology planning is not an easy task. While the path of technological development is often rather predictable, the commercialization of new technology is seldom so. Some technology products become popular on the market, but some do not. Taken together, these difficulties make the formulation of a technology policy a formidable task for most governments. Not surprisingly, in Hong Kong, where the market has traditionally been allowed to direct industrial development, the government has chosen to avoid dealing with technology policy.

In today's competitive global market, every economy has to find its own niche. Setting aside the issue of a technology policy, one can change the angle and ask whether a niche exists for Hong Kong manufacturers, keeping clearly in mind the fact that these manufacturers have a strong manufacturing base in China rather than in Hong Kong. The answer is positive and the niche for Hong Kong manufacturers is in design-intensive products. Both traditional and high technology products may have a high design content. Good design increases the value added of a product and also its profitability. The mass manufacturing of products does not have to take place in Hong Kong; it could take place anywhere. This stage is usually not the one that generates the highest value added. The key to long term growth is to keep the design expertise and the core technologies within the organization. Keeping these core competencies within the organization is important not only because of design-intensive products. In the technology business, the ability to produce incrementally more advanced versions of a product is crucial. One version of the product may fail, but a subsequent, more mature version may be successful. Even if a product succeeds, it still needs to be improved to keep its market share. In either case, a knowledge base is critical in maintaining an edge over competitors.

Technology Products

Today, regardless of the product in question, enhancing it invariably hinges on the use of technology. It is no exaggeration to say that every modern product is a technology product. Technology may have come into play in the design process, in the manufacture of key components, in the choice and testing of materials and so on.

What is really meant by technology and high technology, two equally vague terms, seems to be of some importance to the present discussion. First of all, there is no clear cut definition as to what constitutes high technology. The categorization of high technology products is rather a matter of consensus. Generally speaking, all products utilizing technology resulting from a recent scientific breakthrough, the underlying knowledge of which is still in a phase of rapid progress, are regarded as high technology products. According to this categorization, mobile phones, computers, computer software, video games, disk drives, some kinds of bio-technological drugs, new materials for use in clothing, packaging and electronic components are just a few examples of high technology products. What matters for Hong Kong is probably not whether the product produced belongs to the class of high technology products. So long as Hong Kong manufacturers can utilize technology to bring out enhanced products, a great deal will be gained. The target is technology products in general. Manufacturing high technology products should not be seen as an objective in itself.

There is a demand side and a supply side of any product market and technology products are no exception. On the demand side are the users. Hong Kong users are usually rather quick to catch on with new technology products. Whether they are consumers or producers, they can always be counted on to adopt the latest technology products to further their personal interests or business profits.

The supply side of technology products is more problematic and therefore merits careful analysis. The production of most technology products requires three phases: first, a design phase, second, a manufacturing phase involving the production of the core components and third, an assembly and packaging phase. The relative importance of the three phases obviously differs among different products. For instance, in the case of computer software, the design phase is paramount. In the case of automobiles, the manufacturing phase is just as important as is the design phase. In the case of packaged food, the packaging phase is perhaps the most important of the three.

The word "important" here is used in a sense which gauges the economics notion of value added. For an illustration, take the case of computer software. The value added of a software company is equal to the value of the output minus the value of the material input. Since the value of material input — mainly floppy disks and documentation paper — is almost nil, the value added of a software company is very high — practically the entire value of output. The value created would ultimately be distributed to workers and investors. The payments to workers and investors would then constitute economic income. This income is therefore the contribution of the software company to the economy. In other words, the production of software creates high economic value, which adds to national income.

The value added notion is crucial in the present discussion. It is a matter of strategic importance that Hong Kong manufacturers concentrate on the high value added phases in future manufacturing. Some years ago, many small Hong Kong manufacturers engaged in the "production" of computers. What they actually did was largely to assemble computers using imported components, i.e., the third phase described above. As this was not a high value added phase in the case of computers, the business added little income to Hong Kong. Indeed, it quickly lost out to lower cost countries and disappeared entirely. The same argument can be applied in varying degrees to the assembly of facsimile machines, TV sets, laser disc players and laser printers. Manufacturers typically found that the core components, which were mostly imported, were priced so high that little profit could be made.

The Design-Intensive Niche

If high value added production is so important, the niche in which such products can be produced has to be identified. In the preceding analysis, it was determined that Hong Kong should focus on the design-intensive niche. This niche consists of goods that have a substantial design content. In other words, the value of these goods is basically derived from their design. In terms of the model described above, these are goods which are heavily biased towards the design phase of the production cycle.

What are the typical products in this niche? First, there is the design of more advanced versions of computer hardware. Storage devices such as CD ROMs, hard disc drives and floppy drives are progressing in the direction of higher speed and larger capacities. The evolution of improved versions largely depends on better design. Another area that falls into this niche is application specific integrated circuits (ASICs), which involve the customization of generic chips for specific purposes. This technology makes possible the creation of compact electronic devices that can process information at high speed with the added bonus of using very little power. In Hong Kong, this technology has already made possible the creation of such successful goods as electronic dictionaries and Chinese language pagers.

Second, the design of information engineering software also has very high commercial potential. Software companies such as Microsoft and Netscape are among the fastest growing technology companies in the United States. The key to their success is good, user-friendly product design coupled with clever marketing tactics. The successful development of the G-code for video recorder users is attributable to similar causes. At an abstract level, software enables a better interface to exist between user and hardware. It is becoming increasingly clear that while hardware embodies a great deal of technology, the technology that fills the gap between user and hardware may be even more lucrative. Again, it all boils down to good design.

For software companies, the requirements in terms of physical space and start-up capital are small. There is no reason why Hong Kong cannot be successful in the software business. While the business may require large teams of programmers, the vast supply of high quality programmers in China may be tapped. Manufacturers can conceive of a product idea in Hong Kong, set down its specifications and then commission programmers in China to build the software. Hong Kong can subsequently be used as a test market. If the product turns out to be successful, even larger markets in mainland China and Taiwan can be explored.

Hong Kong also has an advantage in the area of telecommunications infrastructure. The potential in terms of information technology is something that few other cities in the world can match. The trunk lines of the local telephony network are entirely optical fibre, which means that the volume and speed of traffic through the network are exceedingly high. In other words, the network offers strong support for Internet communications. In fact, the number of vendors and users on the Internet have been growing at exponential rates in recent years. Yet, the number of software providers that target the local market is still limited. There is obviously room for substantial growth.

There is yet a third category of design-intensive products that are suitable for development in Hong Kong. There are unlimited opportunities for the creation of new consumer products. The successful experience of the Chinese language pager and the electronic dictionary serves as good inspiration. Less publicized are products like video games, educational toys, electronic pocket games, electronic diaries, smoke detectors, electronic weighing scales, ornamental telephones, cordless phones, consumer medical equipment such as blood pressure and pulse metres and a wide range of household appliances. The key to further growth in this direction is to capture modern, not necessarily frontier, technology and to design it into a product that offers the user superior value. The added value may come from a better choice of materials, better functionality, or simply better looks. The slogan of wisdom in the technology business is that, with good design, even decade old microprocessors can turn a common soft toy into an intelligent interactive companion.[2]

It is not the aim here to provide an exhaustive list of all the technology areas that are suitable for development in Hong Kong.[3] The above list simply presents some examples of design-intensive goods. Indeed, there may well be goods in the biotechnology, materials engineering and environmental engineering fields that have high design contents and are therefore promising areas of development. Surely, even the traditional industries — such the clothing and textile, watches and clocks, printing and jewellery industries — may create higher value by adding more appealing design content into their products.

Intellectual Property Rights Protection

There will always be concern about the fact that design products can be appropriated, and that designers can never capture the full return of their efforts. As Hong Kong is close to China, where the protection of intellectual property rights (IPR) is commonly known to be weak, does the IPR problem loom particularly large?

There is no denying that pirating hurts innovators, but there is little evidence that it destroys creativity entirely. In theory, pirating hurts most in places where a product is sold on a massive scale and all its value is built into the product itself. Music and movie discs are a prime example. The pirating of Hong Kong made CDs and VCDs in China is probably the best known example. Yet, the music industry is still thriving and new firms are starting up. Equally problematic is the pirating of software and databases. Yet, the damage to the inventor may not be as high as one would expect. Many companies insist on procuring the original product for fear of litigation if they do otherwise. Even private users may prefer the original if the purchase includes regular updating. There are also a wide variety of marketing gimmicks that limit the impact of pirating. Moreover, the pirating problem is less serious in the design of some products. For example, electronic components designs, especially if they are intended for a particular customer, do not attract pirating because the sales volume is too small. In the cases of some electronic consumer products, such as electronic dictionaries, the designer has exclusive access to the chip set, which again constitutes a safeguard. We argue, therefore, that the slackness in IPR enforcement does not pose a serious threat to the viability of the design-intensive strategy.

Rationale

The foregoing section suggested a number of reasons for why a design-intensive strategy is appropriate in the Hong Kong context. The rationale of this strategy is summarized in the five points below.

1. The design process is a creative process. Nowadays, production methods are easily standardized. Production according to instructions cannot earn high value added. Many places in the world can carry out this type of production at a very low cost, with no sacrifice in quality. Today, only creativity can earn large rewards.
2. The design process does not require extensive physical

space. It is also non-polluting, an important advantage in a densely populated city like Hong Kong. Moreover, the capital requirement is not large, a feature that small firms would find it to be particularly attractive.

3. If the design-intensive strategy achieves some success, a significant manufacturing presence will remain in Hong Kong. Manufacturers may keep a design centre in Hong Kong. This ensures a continuous generation of income from the goods sector. Whether the goods are manufactured in Hong Kong or elsewhere is irrelevant. Design may be classified as service by the government. This does not matter, because the size of the manufacturing sector, the sector which engages in manufacturing activities, is hardly a social objective in itself. Even though the size of the manufacturing sector is sometimes posed as a government objective by some commentators, this is not an issue to which the government should pay any attention. The reality is that the traditional mode of manufacturing is no longer viable in Hong Kong. No amount of government initiatives is going to reverse Hong Kong's changing comparative advantage.

4. Even a service oriented economy requires a sizable presence of manufacturers, or more accurately, designers. Quality service provision depends on the adaptation and application of technology in the local environment. This process frequently involves close interaction among service providers, hardware suppliers and software designers. Such interaction implies that relying on overseas suppliers and designers does not produce the best results. In Hong Kong, the development of the customer input terminal by Varitronix, a local firm, for the race betting public is a good case in point. (See p. 69)

5. For manufacturers, keeping design centres in Hong Kong has advantages. Hong Kong is more open than is China to the outside world, with efficient air, sea and telecommunication links to all parts of the world. Furthermore, because

Hong Kong people have been widely exposed to both Western and Chinese cultures, their design ideas are easily in tune with consumer tastes. Even though China may have an abundant supply of quality engineers, when it comes to designing a marketable product, Hong Kong talent still has a role to play.

Prerequisites for Success

The crucial question is, what are the necessary factors that would make a design-intensive strategy a success? And does Hong Kong have those factors? To answer the first question, the success of a design-intensive strategy hinges on four factors (Figure 4.1).

1. There must be an adequate pool of knowledge.
2. There is a crucial need for risk taking entrepreneurs who are knowledgeable about market needs.
3. There should be no obstacle to free competition in the product market.
4. There must be adequate financing opportunities for start-up and mature firms.

As for the second question, Hong Kong does have a lot of potential, though it will take some deliberate policies to develop those factors to the territory's best advantage. These issues are picked up in turn in the following sections.

Knowledge Pool

First and foremost, any technology initiative requires an adequate pool of technological knowledge. Since good design invariably involves the use of technology, an adequate knowledge pool is of great importance to the design-intensive strategy. Hong Kong does have a large and rapidly expanding pool of technological knowledge. There are a number of sources of such knowledge. A steady flow of engineers comes out of the local tertiary institutions every

Figure 4.1
Design-Intensive Strategy: Prerequisites for Success

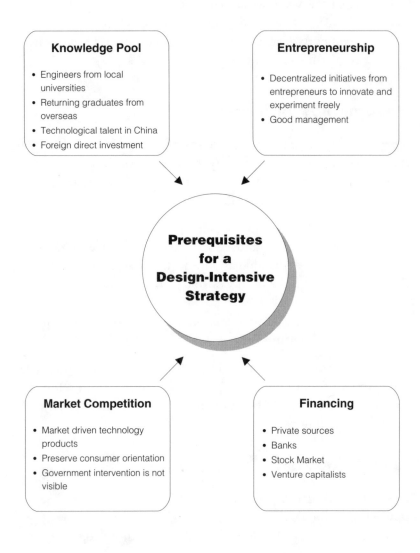

year to join the job market. In addition, there are returning graduates from overseas. Added to this, the huge pool of technological talent in China is gradually becoming a ready source of manpower open to Hong Kong manufacturers. Of course, the manpower from these sources may lack practical experience. In order to develop what these people have learned in school into marketable skills, Hong Kong needs the help of foreign direct investment. Some elaboration on these points may now be in order.

First, local tertiary institutions are producing over a thousand engineering graduates each year. These graduates have received formal training in many different technology fields, such as electronics engineering, information engineering, industrial engineering, automation, environmental science, materials science, biotechnology, chemical engineering and so on. What these graduates need is practical experience. Given the experience, especially if it is provided by world class technology companies, these graduates can make significant contributions to the goods sector.

One often hears the complaint among Hong Kong manufacturers that the best students in Hong Kong are not studying science and engineering. Even those who do choose these fields may decide to become sales executives or programmers and engineers of large utility companies. They have no interest in pursuing a career in technology goods. In response to this complaint, it should be made clear that the problem is not confined to Hong Kong alone. It is a universal problem. Even graduates from top engineering schools in the United States find Wall Street more attractive than Silicon Valley. One also gets the impression that many engineers and research scientists in the United States are not native to that country. There are, in fact, an overwhelming number of Asian scientists doing research in the United States, most of them having received postgraduate training in North America and Europe.

One cannot help drawing some parallels between the United States situation and the one Hong Kong is facing at the moment. In both cases, the best students are in either professional or business programmes. These programmes offer a fast and sure track to the job market. Such jobs offer good salaries and exciting careers. It should come as no surprise that students with the best grades choose to enter these programmes. The problem is that society needs good engineers and good scientists, too. Americans bring in talent from outside. Can Hong Kong do that, as well? Yes, if the government relaxes its policy of admitting foreign students into local tertiary institutions. In particular, if students from China can be admitted into university programmes, both at the undergraduate and graduate levels, then the quality of engineering and science graduates will improve significantly. Of course, there are details that need be ironed out, such as the size of the quota, the fee to be charged, accommodation and financing. But these are not problems so serious that they cannot be resolved.

As for the issue of how engineering graduates can be attracted into technology companies, the root of the problem is in the macro environment. If there are no good engineers willing to work for technology companies, the companies will look elsewhere to set up their operations. With no foreign investment, the environment in Hong Kong cannot be created to attract more investment. Without this environment, it becomes hard to attract engineers to start a career. And the vicious circle continues. To break this circle, the government needs to, in addition to educating better students, take aggressive steps in attracting foreign direct investment. There has to be more co-ordination between the government, the tertiary institutions and the foreign technology companies.

In addition, those engineers who are working in the service sector and in utility companies are engaging in design work. Their experience is not irrelevant to a position in a technology company in the goods sector. Given a favourable environment, they can be attracted into the goods sector and make substantial contributions, as well.

Besides graduates coming out of local tertiary institutions, a second source of manpower is returning graduates from overseas. Over the last few years, a large number of Hong Kong students have gone abroad for further studies. Many have received world class engineering training in the United States, the United Kingdom (U.K.), Canada and Australia. Many of them still maintain strong ties with

Hong Kong and would very much like to work in the territory. In fact, the salary required to secure these young engineers locally is still below that offered for similar jobs in the United States. The manpower cost of setting up operations in Hong Kong relative to setting up in the United States depends very much on the cost of accommodation. A fresh graduate overseas who has a home in Hong Kong would be willing to accept a lower salary here than he or she would in the United States. But for a more senior engineer who has a family in the United States, the salary offered in Hong Kong has to be substantially higher than that he or she would receive in the United States in order to compensate for the high price of housing in Hong Kong. For the overseas investor, the cost of staff alone is not particularly high in Hong Kong if there is an abundant local supply of manpower. If the company has to relocate manpower from overseas or to hire from abroad, the staff cost would be much higher. The implication of all this is that if there is a presence of technology companies in Hong Kong, then returning graduates may not be hard to attract. This points again to the importance of getting technology companies to start up operations in Hong Kong.

Third, foreign direct investment can also be seen as an independent source of knowledge. The way in which technology is utilized, the way goods are designed, the way the market can be read, the way prototypes are tested, and the way products are promoted, priced and distributed can only be learned by participating in these activities. In recent years, Singapore has been much more successful than Hong Kong in attracting foreign direct investment. As a result, Singapore has transformed itself from a beginner in the field into a dominating world supplier of hard disk drives in a matter of years. Although a price has been paid in the process, Singaporean engineers have acquired valuable design and production techniques. These skills will in time be diffused to other companies and industries. Hong Kong was once successful in this regard as well. Surface mounting technology and advanced plastic injection moulding technology were acquired largely through foreign direct investment.

The fourth source of knowledge, which has huge potential, is definitely China. At the moment, government immigration policy does not readily allow engineers from China to work in Hong Kong. The same restriction does not apply to engineers from other countries. There is little reason for such differential treatment. The government seems to be concerned with the appropriate qualifications of the applicants, fearing that allowing Chinese engineers to work in Hong Kong would open a back door for Chinese immigrants of all backgrounds. This worry is probably excessive, particularly if the application has to be lodged through technology companies of established reputation. At the moment, manufacturers have to set up research centres in China in order to tap into China's engineering knowledge. The import restriction in effect benefits neither Hong Kong nor China.

Entrepreneurship

The second prerequisite for a successful design-intensive strategy is the supply of entrepeneurship. Most people in the technology business would agree that the key to creating a successful product is to come up with something that is useful, pleasing, convenient to use and easy to learn. The difficulty is that few people can predict with confidence how well a new product is going to be received by users. Sometimes a superior product does not win the popularity of the market and loses out to an inferior competitor. It is a fact that nobody in the business has the capability of picking winners.

What is most conducive to success from the economy point of view is probably an environment that allows entrepreneurs to innovate and experiment freely. The approach to innovation must emphasize decentralized initiatives from entrepreneurs. Entrepreneurs take the risks of innovation. In the United States, Silicon Valley is probably the most vibrant area of technology research. Saxenian (1994) provides an interesting account of this unique area of technological innovation. A large number of technology companies — from small firms of a few researchers to large ones such as Hewlett Packard and Xerox — are situated in the region. It is often the case that innovation comes from the small firms rather than the large ones, because workers in small firms are more motivated and are allowed more freedom to do research. Yet, the small firms in the Silicon Valley do not operate in isolation. Researchers form networks of their own and help each other with technical problems. Workers also change jobs remarkably often. If a employee believes strongly in an idea but does not get the support of the company in developing it, he or she could start up a new company to pursue it. If it turns out to be a success, the company could grow into a larger company, or it could be sold at a premium. If it turns out to be a failure, the entrepreneur looks for another employer. The flexibility and versatility of Silicon Valley workers lay the foundation for the constant stream of innovations coming out of this area. The lesson to be learned here is that the best arrangement for spurring innovation is one in which a large number of entrepreneurs work in a free and decentralized environment, and in which there is a range of informal networks that allow them to interact easily. Although they belong to a different era, Hong Kong manufacturers in the 1960s and 1970s possessed exactly the same advantages as those which technology firms in Silicon Valley are enjoying now. If the factors that nurture entrepreneurs are still intact in the Hong Kong economy, there is no reason why, given a sufficient knowledge pool, Hong Kong cannot flourish in the technology business.

Many technology firms start off very small, typically with just a few hobbyists. Hewlett Packard, Apple and Netscape are some of the better known examples. What is remarkable about these companies is that they have grown from garage operations into multi-billion-dollar enterprises within years. Motivation, ideas and knowledge are not sufficient to achieve success of this scale. A big part of the success story is good management. These companies needed a manager to arrange financing, organize workers, lay out production plans, market products, prepare financial statements and look after the legal aspects of the business. The point is that the technical side and the management side are equally important in building up a technology business. If a technology company has both good management and good engineers, whether or not it has a winning product is not of primary importance. Such a company will be able to create new and better products to replace the old ones and bring about long term growth. With only one good product, any success is going to be short lived.

Entrepreneurship and management are Hong Kong's strengths. These strengths have been nurtured through years of hard earned experience. Although the technical aspects of running businesses in the technology age may be different from what they once were, the basic principles of taking market risks and managing people remain largely unchanged. The basic ingredients for success are already in place. What appears to be missing is a mechanism which enlarges the knowledge pool and improve the public's awareness of modern technology.

Market Competition

Some technology products are technology driven. The evolution of technology follows a predictable path along which products of one generation lead to those of another. Even if the development of certain products is not expected to generate any profits, there are always some companies that would find it worthwhile to pursue. These companies may be supported by their governments or from earnings derived from other products. The arena of technology driven products is not one that Hong Kong should enter, however.

In the design-intensive niche, technology products are largely market driven, not technology driven. Users decide which of these products is best. It is important to preserve this consumer orientation. The consumer is not concerned about which product represents the higher technology level. He or she is concerned only with the utility of the product, its costs and its appearance. It is therefore absurd for the government to try to subsidize or protect certain products. Any such intervention would only stifle the development of competing products and slow down the product cycle. The result would be a weakening of the product's ability to enter foreign markets. In an economy like Hong Kong's, which relies on exports for long term growth, subsidy and protection of individual products can never be a profitable proposition.

In this regard, it may be worthwhile to evaluate a suggestion that asks the government to move in the direction of import substitution. It was suggested that the government should direct the development of industries to replace the importation of certain key components that are being used by Hong Kong manufacturers, i.e., memory chips, microprocessors, laser heads, watch movements, TV screens and so on. The prices of these locally made components would be lower, according to the suggestion, and this would help Hong Kong manufacturers become more competitive.

The problem with the suggestion is that the creation of local industries with an aim to substitute for imports is generally not advisable. There are usually good reasons for the absence of industries which produce certain components. Maybe the start-up capital is very large, or the technology is controlled by a small number of firms. Asking the government to create these industries works against the comparative advantages of Hong Kong. Moreover, these industries will, by objective, be inward rather than outward oriented. The objective of producing for local use could easily lead to inferior quality and backwardness in design. It will turn out that import substitution does not really increase competitiveness. It should be emphasized that there is a difference between the government creating an industry for import substitution and the government attracting foreign investors who produce components that are supplied to Hong Kong manufacturers. Foreign investors keep uniform standards across all their manufacturing sites and their products will have to compete in the world market. Moreover, their choice of products and production methods have to be compatible with the comparative advantages of Hong Kong.

The government should be supporting generic technologies. These are technologies which can be applied to a variety of products. Even though companies that receive such support may not turn out to be profitable ventures, the technology developed adds to the knowledge pool of Hong Kong and the human resources stay within the economy. Their knowledge may diffuse over time and bring about the creation of products of a different form and nature.

Relative to other newly industrialized economies, Hong Kong

enjoys an advantage in that it has a large local market that is receptive to new technology products. For example, the Singaporean market, which is highly educated, is only about half the size of that of Hong Kong. The size of the local market allows Hong Kong to play a valuable role as a test market of new products. If a product sells well in Hong Kong, the next step may be to enter China. The cultural proximity of Hong Kong, Taiwan and the PRC induces manufacturers to view them as one big market. The presence of Hong Kong as a test market is of great advantage to manufacturers. If a product meets with some local success, the manufacturer gains not only the confidence to explore other markets, but also earnings that support further research. In some sense, a R&D budget is decided upon in much the same way as an advertising budget is. If a new product sells well initially, more revenue will be available for R&D, which in turn further improves the product.

Financing

Financing is crucial to every stage of the development of a firm. As Hong Kong is rich in investment capital, but has a limited number of sources that can finance technology firms, the financing aspect of technology development merits a detailed discussion. Generally speaking, a firm derives its funds from one of the following sources: private sources, banks, the stock market and venture capitalists. These are discussed in turn below.

Private

Probably the largest source of financing is the private source. This is true not only in Hong Kong, but in most other free economies. The private source consists of the retained earnings of the business, the personal savings of the owners, or personal loans extended to the owners. As the users of the funds are the same persons as, or are very close to, the providers of the funds, the cost of effecting such financing is usually the lowest. The chance that the funds will be misused or improperly allocated is also small. In fact, many technology companies rely solely on private financing. The owners keep the business under their full control, and, since they borrow little, their business will not be put under stress if future earnings turn out to be low.

Of course, a great deal of research has to be done for technology companies to stay ahead of peers, or simply to keep abreast of them. In most cases, funding for research comes from private sources, even though the research budget may be huge. For technology companies, the industry norm is to invest about 10 per cent of sales revenue into researching new products.[4] The research process takes place on a continuous basis and results come incrementally. When business is good, revenue increases, more is put into research, new products come out more quickly and the product life cycle shortens. Successful companies grow very fast, but failing firms also die quickly.

Banks

In a capitalist economy, banks are the major financial intermediaries. Banks channel the savings of depositors to finance investment ventures. Because banks are in effect investing depositors' money, they are primarily concerned with the repayment capability of the borrower. The return to the banks' investment is basically in the form of interest payment. Even if the borrower does extremely well, banks do not have a share in the profits. The upside potential of their investments is therefore limited. Yet, banks stand to lose their entire investment if the borrower defaults. To minimize their downside exposure, banks usually require collateral from the borrower. Not all assets qualify as collateral. In Hong Kong, the two main types of collateral are real property and cash deposits. Young technology firms usually have no such assets and therefore support from banks is hard to obtain.

Even though the lack of bank support may be seen as a handicap to young technology firms, some firms think otherwise. To them, over-expansion is just as great a danger as is stagnation. Expansion through borrowing reduces a company's ability to survive hard

times. To a technology company, resilience is of particular importance, because the technology business is subject to extreme ups and
downs.

The Stock Market

The stock market is also an important financial intermediary. The
public can invest its money in a company's shares and share in the
company's earnings and growth. From the company's point of
view, management receives outside funding without giving up control of the company or increasing the burden with interest payments.

One attractive advantage of stock market investment is liquidity. Because a company's shares are publicly traded, investors know
they can liquidate their investment or adjust their portfolios easily.
For a more subtle reason, the liquidity of public share trading is also
a significant impetus driving some technology companies' decision
to go for public listing. The reason has to do with the organizational
structure of technology companies. Because of the knowledge intensive nature of these companies, upper management has difficulty
monitoring worker efforts. It is far easier for upper management to
motivate than to monitor workers. One way to provide motivation
is to give workers shares in the company. The best way to provide
workers with liquidity is to allow public trading of shares.

Another advantage of public listing is that it allows a company
to acquire other companies without having to pay cash. This advantage is more applicable to the United States than it is to Hong Kong,
as there are far more mergers and acquisitions in the United States.
In the United States, many technology companies have to go into
unfamiliar lines of business to continue growing, and an easy way to
obtain financing is by exchanging shares.

Not all companies qualify for listing on a stock exchange. In
Hong Kong, the Hong Kong Stock Exchange sets down rules with
regard to the financial record and soundness of the company and requires the sponsorship of investment banks in underwriting the new
shares. The purpose is to protect the public from unscrupulous

companies. Of course, the regulations keep away the good as well as the bad. For small to medium sized companies with no established track record, the stock market is generally not accessible.

Venture Capitalists

Venture capitalists are an important source of financing in today's technology business. In the United States there are 700 to 800 venture capitalists, most of whom invest in technology firms. A venture capitalist is basically a professional investor who amasses funds from large institutional investors and channels them into equity investment in fast growing companies. Commensurate with the higher risks involved, they look for a return to investments of at least 25 per cent per annum. The period of investment is three to five years. In order to be able to liquidate their investment at the end of the investment period, venture capitalists require a clear exit route, which may take the form of a public listing or a sell-out to another company. In this regard, the existence of a smoothly running stock exchange is a major advantage in the eyes of a venture capitalist.

Before making investments, venture capitalists screen firms carefully. They investigate a firm's management quality, its financial statements, the market conditions and the exit route. They accept only a small fraction of the proposals they receive. The rejected firms learn a lesson, repackage their proposals and approach other venture capitalists. When venture capitalists do invest, they usually offer a firm much more than just money. They help make strategic plans, manage the company, arrange for other sources of financing and advise the company on public listing procedures. To technology firms inexperienced in the business world, venture capitalists are an invaluable source of assistance.

In Hong Kong a sizeable presence of venture capitalists has existed for some time. By the end of 1994 there were already 73 funds representing capital of US$6,037 million. These funds targeted the Hong Kong and China region. In addition, there were another 39 funds representing capital of US$2,384 million targeting China alone. What is interesting is that few of them have invested in technology companies in Hong Kong.[5] The reason for venture capitalists' lack of interest in technology investment is unclear. One possibility is that Hong Kong has not been known for its achievement in technology development. There are not many local technology companies of sufficient merit to warrant the attention of venture capitalists. The funding is there, but venture capitalists look elsewhere in the region — to Singapore, Malaysia and China.

This concludes the discussion of the four prerequisites necessary for the success of a design-intensive strategy. On all four counts, Hong Kong has potential. Hong Kong has a growing pool of trained engineers, and this pool will become even larger if China's vast potential can be tapped. Hong Kong has a long tradition of nurturing entrepreneurs, and it possesses an environment that encourages them to interact closely with each other. Hong Kong has a freely competitive market for products and the size of the market makes it an ideal testing ground for new products. Finally, Hong Kong is rich in investment capital. Its financial centre status also attracts funding from foreign sources. However, there has been a lack of foreign direct investment in recent years, leading to a lack of practical experience among young engineers. The level of technology awareness is also rather low. There is as yet no widespread interest in investing in technological ventures. More has to be done to create a better mechanism by which funding can be channelled into technological ventures. These considerations lead to a host of policy issues, which will be picked up in Chapter 6.

Having set up a framework to address the strategic issues of technology and industry, a number of controversial arguments may be analyzed, including the role of the universities, the technology lag, and what to do with the traditional OEM products.

The Role of the Universities

It was mentioned earlier that technology can be seen as a knowledge base. As the design-intensive strategy emphasizes the adaptation of existing technology to new uses through better design, the strategy's most urgent requirement is the accelerated transfer of technology. Is this task an educational one that should be left to universities?

The first part of the argument, that the design-intensive strategy emphasizes technology adaptation is completely in line with what has been discussed so far. But identifying technology transfer as an educational process that should be left to universities is somewhat problematic. Universities play an important role in the diffusion of generic technological knowledge. Nevertheless, in the design of study programmes, institutions are typically technology driven rather than market driven or product driven. There is a good reason for being technology driven. Technology evolves in a predictable manner, whereas its use in the creation of new products is volatile. Universities do not teach things that become obsolete within a few years. They teach students so that these students will be able to fit into a wide range of settings and to learn by themselves through practical experience. Therefore, good higher education is essential, but it is only the first half of the story. The second half is in the placement of students in a setting in which they can continue learning through on-the-job training. This is where they apply their knowledge to making market-driven products. Practical experience should come from working for well established technology companies. In the case of Hong Kong, this would mean working for foreign direct investors.

A second argument concerns whether local universities can play a more active role in the commercialization of technology. Inspiration for this argument probably came from Stanford University's success in spurring on the development of Silicon Valley in California. Yet, the Stanford example is probably an isolated case. It is hard to find an equally successful example elsewhere in the United States, or even in the world. Indeed, Harvard and MIT have not enjoyed the same success in the development of Route 128 ventures as Stanford has with Silicon Valley. The collaboration between university and industry is something that should be allowed to evolve by itself. Planned directives do not work in most cases.

The primary aim of university research is to extend the knowledge frontier, not to make commercial profits. Of course, there are areas in which new knowledge can be quickly utilized in improving commercial products. But even in those areas, university researchers seldom cross the line to develop their ideas into a product prototype and further into a marketable product. The process requires far more human and financial resource than most university researchers would bother to amass. It may be more productive to view universities as a repository of research findings which commercial firms may tap. University researchers and commercial firms may then be able to find suitable venues in which to undertake cooperative research. This type of collaborative research may also result in undergraduate and graduate students joining commercial ventures and starting their careers in the industries. In the U.K., a "Teaching Company Scheme" brings postgraduate students in to conduct research on specific problems for outside companies. The scheme promotes students' awareness and increases their market value upon graduation. This is an area worthy of further study in the Hong Kong context.

The Catch Up Game

One often hears the argument that Hong Kong is small and devoid of technological personnel. Relative to the United States, Japan, Germany and other developed countries, Hong Kong is so far behind in terms of technology that it is never going to catch up. Hong Kong should, so goes the argument, follow the trend and let the economy move into finance, trading and business services. Investing in technology is going against the comparative advantage of Hong Kong and is therefore a waste of resources.

At the moment, the United States and Japan are the world's two leaders in technology. The United States is leading in the areas of nuclear technology, aerospace, biotechnology, software and pharmaceuticals. Japan leads in the areas of microelectronics, photoelectronics and robotics. China is also very advanced in the areas of microelectronics, nuclear engineering, aerospace and biotechnology, though engineers in China lack sufficient exposure to product markets to make commercializing their skills easy. Both the United States and Japan are investing enormous amounts in technology research to maintain their leading positions. As of now, advancement in research has outpaced its commercial application. Business firms are in some cases holding back on commercializing new technology to allow a gradual phasing out of existing products. As a result, products that are being put on the market today do not actually incorporate the newest technologies.

Hong Kong should not aim at competing against technology giants on those fronts. It cannot catch up by taking their products and doing reverse engineering. Furthermore, the manufacturing of many technology products, such as microprocessors and RAM chips, is subject to enormous economies of scale. As production volume grows, the unit cost of products drops continuously over a large output range. Even large manufacturers are undercutting each other's prices and having problems keeping their profit margins. Hong Kong has no comparative advantage in manufacturing such products. However, it does have an advantage in producing design-intensive goods. These goods may be designed using existing technology in an innovative way. Alternatively, the goods may require custom design to suit a particular purpose. The giants may have monopolized the technology frontier, but there are still niche markets open to Hong Kong.

One should not conclude, however, that no basic research should take place in Hong Kong. In some technology areas, the ability to utilize existing technology requires basic research. The conceptual difference lies between basic research with an aim towards scientific discovery and basic research intended to enhance technology transfer. It is in the latter category that Hong Kong is in need of improvement.

Traditional Products

A final point that needs to be addressed in the present framework is whether it is necessary for Hong Kong manufacturers to climb the technology ladder. Some people believe that Hong Kong manufacturers need not do technology research. If they can produce better products at a price lower than that of competitors, they can already

make immense profits. Traditional goods such as watches, spectacle frames and costume jewellery could be produced very
profitably. The risk involved, they believe, is probably lower than
that associated with technology research.

The argument had some validity in the past, but this is no longer
the case. Many products, such as athletic shoes, clothing and soft
toys are actually getting cheaper despite consumers' increasing income. The decline in prices has come about largely as a result of the
release of manpower from liberalizing countries. There is no way
that a Hong Kong manufacturer can compete in these businesses
unless it moves from being an OEM manufacturer to being a product designer. True, manufacturers can move further inland into
China and tap the cheap labour there. But the feasibility of such action depends heavily on the support of telecommunications, energy,
transport and port infrastructure.

Moving up from OEM to product design invariably involves
technology. It is clear that even traditional products such as clothing, athletic shoes and soft toys require a great deal of help from
technology. As an example, consider clothing production in the
past and the present. The machining of the articles of clothing is
probably the only part of the process that has not undergone significant change. Nowadays, clothing patterns are electronically stored
in computers, price quotations are produced by software, designs
are created by computers, components are cut by laser, materials
come from laboratory synthetics, and the production processes are
co-ordinated via fibre optics and satellites.

It is true that manufacturing some traditional products can be
profitable for a while. Nevertheless, such activity rarely brings
about sustained corporate growth. Because Hong Kong is such a
fertile ground for entrepreneurs, examples of success easily capture
the public's attention. Few people realize that many such firms
quickly plateau in the absence of new product initiatives. Furthermore, this kind of profit opportunity is becoming scarce. In the
future, while Hong Kong manufacturers may continue their usual
quick-profit strategy with a few traditional products, more technology products need to be developed to support higher wage rates and

input costs.

Notes

1. See Simon (1995) for studies on technology policies in countries of the Pacific Rim. For studies on Japan, see Johnson (1982) and Okimoto (1989).

2. For an inspiring account, see Kao (1991).

3. See Kao and Young (1991) for a discussion on the strategic technology areas suitable for development in Hong Kong.

4. For a reference on how people in the industry look at research policy, see Gomory (1992).

5. Figures extracted from *Asian Venture Capitalist Yearbook 1994*.

CHAPTER 5

The Design-Intensive Strategy in Action

It has been argued throughout this monograph that the production of goods is important to the future growth of Hong Kong, that Hong Kong possesses the competitive advantages necessary to the pursuit of a design-intensive strategy, and that technology has found its way into improving all kinds of products. In many ways, Hong Kong manufacturers have already begun to implement a design-intensive strategy. New and successful products have emerged in the forms of liquid crystal displays, electronic dictionaries, electronic games, clothing, jewellery, food and beverage, packaging, printing and publishing, and fashion items such as toys, watches and spectacle frames. All these products illustrate the application of design efforts in combination with modern technology to produce a good with substantial value added. The value added margin ultimately accrues to the workers and the entrepreneurs who find the market niche. The design-intensive market niche is not a dream; it is a reality to some manufacturers. There are some well documented cases of companies that grew from start-up firms to publicly listed companies within a few years. Their success stories effectively illustrate the benefits that a design-intensive technology strategy could bring to the Hong Kong community at large.

Liquid Crystal Display

The manufacturing of liquid crystal display (LCD) units is a good case in point. LCD units are display devices commonly used in

67

watches, calculators, electronic diaries and dictionaries, hand-held computer games, mobile phones, pagers, laptop computers and many kinds of machinery. The process of manufacturing LCDs is very labour intensive. It includes five stages. The first stage involves preparing a thin glass surface. The second stage involves the photolithography of electrode patterns to form character patterns. The third stage involves injecting fluid into an envelope made from two pieces of glass and the sealing of the envelope. The final stage involves the fitting of the electrodes. The whole process must take place under exacting standards, in an environment requiring strict control over air quality and electricity supply. Machine automation is not considered practical, as much of the manufacturing process requires human judgment and a great deal of skill.

Varitronix, established in September 1978, is a technology company that designs, manufactures and sells LCD units.[1] The manufacturing processes take place mainly in China, while designing is done in both China and Hong Kong. The world's largest producer of LCDs is Japan. Varitronix knew early on that mass production of standard LCDs would only bring the company into direct competition with large Japanese manufacturers. That strategy would hardly be a viable proposition. In 1981 the company found its market niche. It began to custom design LCDs for equipment manufacturers. The company was particularly strong in fulfilling small lot orders and in making prototypes. Over the years, the mass production of standard LCDs has been losing importance for Varitronix in the face of the production of high value added custom designed products. By 1990 mass volume products only accounted for 38 per cent of company sales.

Like many other manufacturers, Varitronix maintains two production facilities, one in Hong Kong and another much bigger one in south China. The China facility, established in 1983, has grown rapidly since its inception. By 1990 the company had completed a seven storey factory in Sha Wan, near Shenzhen, with 66,000 square feet of factory space.

In recent years, the company has diversified and it now makes self-functioning electronic components, such as those which form the heart of electronic spellers and dictionaries. The creation of these components, liquid crystal modules (called LCMs), requires the mounting of integrated circuits and LCDs onto a printed circuit board using surface mounting technology. Varitronix is capable of offering equipment manufacturers one-stop shopping advantages. It can design modules, source integrated circuits and printed circuit boards, and manufacture complete LCMs. Its design capability, together with the one-stop shopping advantage, have enabled it to secure a viable market niche.

A good example of this niche market is the customer input terminal (CIT) which the company developed for the Hong Kong Jockey Club. The customer input terminal, which is patented worldwide, is a hand-held, palm size device. In its current application, the CIT has a large LCD unit which is touch sensitive. The unit can display both English and Chinese characters. The user can use the CIT to place bets and receive horse race results anywhere in Hong Kong. The technology of the CIT has wide ranging potentials. For example, it could be adapted for use in home banking.

The development of the CIT has won the company long term supply contracts from the Jockey Club. The presence of a long term buyer provides a platform upon which the company can develop more advanced generations. The company's capability in the technologies of the LCD, integrated circuits, printed circuit board and surface mounting means that it can research, design, prototype and manufacture more advanced products.

The financial performance of Varitronix has been nothing short of spectacular. The company was founded in September 1978 with an initial investment of HK$2 million. Production started in January 1979, and the company has been profitable ever since. By 1986 it already had a business turnover of HK$85 million and a profit after tax of HK$26 million. By 1990 turnover had shot up to HK$229 million and profit after tax to HK$69 million. The company became publicly listed in June 1991. At the time of the listing, the company was worth HK$682 million, some 340 times its starting value in 1978.

A knowledge pool is always the most important element behind the success of any technology company and Varitronix is no exception. Probably the most important driving force behind its success has been the human resources the company possessed from the very beginning. The two key founders of the company are themselves technologists. C. C. Chang is a doctoral graduate of the University of Manchester Institute of Science and Technology in solid state electronics and York Liao is a doctoral graduate of Harvard University in applied physics. Before founding Varitronix, both were faculty members in the Electronics Department of The Chinese University of Hong Kong. After returning to Hong Kong to teach at the university, Liao established connections with Fairchild Semiconductor, an American company that at the time was the only LCD manufacturer in Hong Kong. Liao became aware of the practical difficulties with, as well as of the opportunities in, LCD manufacturing. At the university, he set up a small laboratory for LCD manufacturing and trained a number of graduate students and technicians. Some of these later joined Varitronix when the company first started up.

Chang and Liao bring an ideal mix of talents to a technology company. Chang is a good administrator with a great deal of entrepreneurial drive. Liao is a scientist with deep technological knowledge. Together they have solved many production problems that have saved the company money and kept it profitable in a competitive environment.

The success of Varitronix is a good illustration of a number of arguments made in the previous chapter.

1. Design-intensive niche — Varitronix is successful primarily because it has targeted the design-intensive niche. The design content of its products yields very high value added. In the case of many models, completing the prototypes already allowed the company to break even.
2. Human resources — Varitronix has both entrepreneurial skills and scientific knowledge, a combination which is critical to any technology company's success. The technology was acquired from two sources. The scientific

knowledge was acquired when the founders received their training overseas. More importantly, the practical production and market experience was acquired through their connection with a major foreign owned LCD manufacturer. The technology was solidly in place before the company started operations.

3. Financing from retained earnings — Varitronix does not have a separate R&D department. Research occurs as part of the ongoing process of new product development. Hence, there is no R&D budget. The company has been very conservative in terms of its financial management. It has avoided incurring long term debts and growth has been financed primarily by retained earnings. It did go for public listing on the stock exchange, probably not for financing reasons, but for a fair valuation of the company. The company is fortunate in that it had adequate cash flow from the very beginning. The cash flow helped support the development of new products and maintain a sizable staff of technicians.

Electronic Dictionary

The design of components such as LCDs is not the only way a design-intensive strategy can proceed. The design of innovative consumer products could make a big hit within a very short time. The example of the Chinese electronic dictionary provides a good illustration. It is a consumer product that did not exist a decade ago. The major driving force behind this product is a young company by the name of Group Sense.[2]

Group Sense is a company that designs, develops, manufactures and sells a range of electronic consumer products under its own brand name. Its major product line is electronic Chinese dictionaries, which are supplemented by a line of multilingual translators. For the 1992 financial year, electronic dictionaries accounted for 62 per cent of company turnover and translators accounted for 19 per cent. About half of the company's products were exported. In the local market for electronic personal Chinese dictionaries, the company had a market share of 45 per cent.

As a start-up company, Group Sense grew very quickly. In June 1988 two entrepreneurial youths, Samson Tam and his brother, Thomas Tam, founded a company that provided engineering and consultancy services for electronic toys to local manufacturers. Samson Tam is a degree holder in electronic engineering from The Chinese University of Hong Kong and Thomas Tam is a diploma holder in electronic engineering from the then Hong Kong Polytechnic. The two young founders, then in their early twenties, saw the market potential for electronic personal dictionaries. That was an era during which China was quickly opening to the world and many Chinese were eager to learn English. Using their experience in designing electronic toys, the Tams quickly developed the prototype of their first Chinese dictionary model. At that time, they had no manufacturing facilities, but due to the presence of linkage support provided by a large range of component manufacturers, they managed to arrange a joint venture to launch their first product on the market. The product was an instant success. Spurred on by this success, the company built a large production facility in August 1990 in Shenzhen, China, to cater to the demand for its products. Its Hong Kong office was basically used as a research and development and administrative support centre. From then on, almost all mass manufacturing took place in the China facility which by 1992 had grown into a complex of three multi-storeyed buildings accommodating more than 800 workers.

The key to the success of Group Sense is superior design and quick development of new products. The core components of its products are microprocessors and Mask ROM chips. The design of these components involves application specific integrated circuits (ASIC) technology. This technology, which is also used in electronic games, builds software into hardware permanently to achieve high computation efficiency.

The company's financial performance has been astounding. In the year ending on 31 March 1990, just two years after it was founded, the company had already recorded a turnover of HK$38 million and a profit of HK$6 million. For the financial year ending in March 1992, turnover shot up to HK$157 million and profit to HK$49 million. In early 1993 Group Sense became a publicly listed company on the Hong Kong Stock Exchange. At the time of listing, the five year old company was worth, based on market capitalization, HK$800 million.

The electronic dictionary market is very competitive and is completely user driven. A number of local and Taiwanese manufacturers have produced competing products. Competition forces manufacturers to develop products fast and the reward is a quickly expanding market in the PRC, Hong Kong and Taiwan. To speed up the development cycle, Group Sense explored and gained access to research results around the world. Besides maintaining a design team in Hong Kong, the company had access to the output of a research centre in Shenzhen. Sci-Sense, a research centre employing fifty engineers, was established as a joint venture by Group Sense, Shenzhen Technology Research, and Shenzhen Sunshine Technology. An advanced model of the dictionary launched in 1992 could translate complete sentences from English to Chinese. The technology used in this model was provided by software developed by the Institute of Computing Technology, Chinese Academy of Science, which agreed to license the software to Sci-Sense. Market pressure for new product generation also induced Group Sense to license technology from overseas. To enable the company to improve voice generation in some dictionary models, it licensed text-to-speech programmes for foreign languages from BST. In addition, the company licensed foreign language dictionary databases from Houghton Mifflin in the United States.

The case of Group Sense effectively illustrates a number of points made earlier. First of all, while none of the technologies used by the company were really cutting edge, it has proved that clever commercialization of existing technology to cater to a large regional market can produce large profits. Of course, even though the development of cutting edge technology is not essential, keeping abreast of technology development worldwide is important. In this

regard, Group Sense has made use of a number of strategic alliances
with research centres in China, licensers and technologists in the
United States, and chip manufacturers in Japan to help develop new
models rapidly. Group Sense has also made good use of researchers,
mainly software engineers in China, to develop its more powerful
models. The technology needed to translate English sentences into
Chinese has been in existence for some time. Sci-Sense captured it
through joint ventures in China and commercialized it successfully.
In addition, the pressure to innovate originated from keen market
competition. Local and Taiwanese manufacturers had developed ri-
val dictionary models. Free competition with no government
regulation or protection fostered rapid product improvement, and
with the improved models Group Sense was able to gradually shift
its market overseas. Finally, the electronic dictionary industry was
not created in isolation. It received support from many local manufacturers who supplied components such as circuit boards, plastic
casings and liquid crystal displays. Without the support and knowledge of these linkage industries, the company's entrepreneurial
innovation might never have turned into reality.

Chinese Pager

The Hong Kong community by nature requires many telecommunication devices. Mobile phones, pagers and facsimile machines in
Hong Kong have one of the highest penetration rates in the world.
One in seven persons in Hong Kong is a mobile phone user and one
in six is a pager user. The development of the pager is a good illustration of the way in which technology can enhance telecommunication services.

The pager is a lightweight, wide coverage and inexpensive device. The early generation of pagers could only transmit a "beep"
tone to the user, who then had to call the system station to retrieve
the message left by the caller. A subsequent generation allowed a
code to be transmitted to the user, who could tell by the code the
category of message left by the caller. This was followed by a much
more powerful pager version which allowed messages in English (or

other languages in alphabets) and numbers to be transmitted and displayed directly.

The English language pager is evidently a valuable technology product in English speaking countries. It is, however, not ideal in the Hong Kong context, as callers and service subscribers in the territory are predominantly Chinese. Station operators originally had to translate messages from Chinese to English before they could be transmitted to users, so translation errors could easily arise. There was a clear need for a Chinese language pager. A company by the name of Champion Technology subsequently emerged to fill this gap.[3] Champion Technology appeared to be the first company to successfully develop a Chinese language pager. According to a company listing prospectus, it developed not only the pager, but also the hardware and software of the paging system that enabled a complete provision of Chinese paging service.

The backbone of Champion Technology's management team was made up of former employees of Asiadata, a subsidiary of Cable and Wireless. Cable and Wireless was the majority shareholder of Hongkong Telecom. Paul Kan, president, who founded the company in 1987, had worked for Asiadata for fifteen years. James Carter, who joined Champion Technology as managing director in 1992, had also been at Asiadata for fifteen years. Eric Poon, vice president, was another long time employee of Asiadata.

Champion Technology's management gained much more than just experience from Hongkong Telecom. The close connection between Champion Technology and Hongkong Telecom fostered the formation of strategic alliances between the two companies. When Champion Technology succeeded in developing the world's first multi-lingual pager and paging system in 1988, it had neither a paging license from the government, nor a transmission system. Therefore, it licensed the system to Hongkong Telecom, which set up the transmission network and started the new service in 1989. The move ensured an early return to the company's research investment.

The company did little manufacturing. Its Hong Kong plant basically engaged in the final assembly and testing of pagers. The manufacturing of the pagers' components was undertaken in Japan using materials manufactured in Japan and Hong Kong.

Champion Technology has an illustrious financial record. According to its listing prospectus, in 1990 the company made a profit of HK$1.8 million. In 1991, as a result of extensive licensing to China and Taiwan, profits jumped to HK$23.7 million. In 1992, as a result of joint ventures in China to set up paging systems and operations, profits soared to HK$80 million. The company was publicly listed on the Hong Kong Stock Exchange in August 1992. At the time of listing the company was worth HK$600 million.

The example of Champion Technology illustrates the way in which existing technology can be adapted to the local environment through design-intensive work. The company found a niche for Chinese language paging, exploited the latent demand and created a new industry. With the boom in telecommunication products worldwide, opportunities for new products abound. There is a consensus among technologists that despite technological advancement, there are certainly gaps in the interface between technology and users. The challenge is to bring technology to consumers without requiring them to learn and adapt. These gaps need to be filled if technology products are to be user friendly. Some of the gaps are specific to communities in the region because of these communities' languages and cultures. In this regard, the development of the Chinese pager met an important need. Creating the pagers might not have required high technology in the strict sense. The important thing is that the company used technology creatively to serve users and make profits. It is profit that drives entrepreneurs, not high technology.

These are just a few examples of innovations that Hong Kong entrepreneurs have successfully launched in the past. There are undoubtedly many more such cases, though they are not as well documented. Studying these cases, one gets the general impression that Hong Kong has advantages in developing products that are geared towards the consumer. The technology required to develop the products discussed above was not excessively demanding.

Table 5.1
Summary of Case Studies

VARIABLES	COMPANY NAMES		
	Case 1 Varitronix	Case 2 Group Sense	Case 3 Champion Technology
Study Period	1978–1991	1988–1993	1987–1992
Major Product	Liquid crystal displays	Electronic dictionaries	Design of Chinese pagers
Product Nature	Hardware component	Consumer product	Telecom product
Manufacturing Plants	Mainly in Shenzhen	Mainly in Shenzhen	Contracted out
Design Centre	Hong Kong and the PRC	Hong Kong and the PRC	Hong Kong
Design Content	High	High	High
Financing before Listing	Private	Private	Private
Years to Listing	13	5	5
Market Capitalization at Listing	HK$682 million	HK$800 million	HK$600 million
Knowledge Source	Founders	Founders	Founders
Entrepreneurship	Strong	Strong	Strong
Licensing of Outside Technology	Minor	Major	not applicable (just assembling and testing)
Export Volume	High	Moderate	Low

Nevertheless, the four ingredients for success — namely, a knowl-edge base, entrepreneurial spirit, product improvement through free competition and adequate financing — were in place for the companies that created them. Other firms may not be as fortunate. At the economy level, there is a need for mechanisms that facilitate timely access to these four ingredients by the entrepreneurs. This is the way to ensure the long term viability of the goods sector and also the growth of the Hong Kong economy. To achieve this goal will be a major challenge requiring government leadership.

Notes

1. Corporate information summarized from Kao and Young (1991) and company listing prospectus.

2. Corporate information summarized from company listing prospectus in 1993.

3. Corporate information summarized from company listing prospectus in 1992.

CHAPTER 6

Policy Considerations

Overview

In Hong Kong, there is no history of government direction in industrial development. There is effectively no industrial or technology policy. Initiatives to develop new products and create new industries have always come from entrepreneurs in a decentralized fashion. The government has viewed its role as one of providing a suitable environment for industry development. This refers primarily to the governmental task of ensuring the adequate provision of infrastructure — seaport, airport, roads, water, power and so on. The plans for building infrastructure are not drawn out with the express aim of serving industries, but are rather geared towards supporting all economic activities in Hong Kong. The planning and execution of policies related to infrastructure come under the purview of various government departments. The Industry Department, however, is not one of them. In Hong Kong, not only is the composition of the manufacturing sector left to market forces, but the weight of the manufacturing sector relative to that of the service sector is, too. Market forces determine the sizes of both sectors.

The issue at hand is whether in the face of rapid technological development worldwide, some fundamental changes in government philosophy are required. It was argued in Chapter 1 that the goods sector is an important driving force for growth, owing to the export potential of the sector. In Chapter 4 it was argued that the goods sector should focus on the design-intensive niche, and that the implementation of a design-intensive strategy hinges on the development of technology. Therefore, while it has been maintained that the government should avoid direct intervention in the product markets, the government should provide leadership in the development of technology. Technology should be treated as the knowledge base that supports economic growth in general.

There are a number of areas in which the government can assume a strong leadership role. It must first increase the level of its own technological expertise. The Industry Department should employ full time technologists. A high level advisory committee should be established to review the state of technology in Hong Kong and to guide the Industry Department in implementing technology initiatives. The government's second step should be to strengthen the knowledge base in Hong Kong by allowing selective importation of engineers from China. Its third step should be to take aggressive steps to attract foreign direct investment.

These three steps have to be taken rather quickly in order to create an environment that is attractive to technology companies and young engineers. Additional steps can be taken to enhance the public's awareness of technology and to improve the financing mechanisms of technology companies. Ultimately, the technology transferred into Hong Kong has to be adapted to the local context. In time indigenous research has to be developed too. Knowledge acquired must be diffused to impact a wider population, so that new products that target the local market as well as those in other regions of the world can be created. The ultimate objective of the design-intensive strategy is to use technology to create products that are uniquely suited to a market niche. A summary of the policy suggestions is presented in Table 6.1.

Technological Expertise within the Government

Hong Kong is generally perceived to be lagging behind other countries in technology. This perception gives rise to a vicious circle. Foreign technology companies do not choose Hong Kong as a manufacturing or design centre. In the absence of these operations, young engineers are deprived of the practical experience they need. Seeing the lack of prospects, students do not want to become engi-

Table 6.1

Summary of Policy Suggestions

Action	Time	Objectives	Remarks
More technologists in Industry Dept.	Immediate	Boost government expertise	
Hi-level advisory committee	Immediate	Steer government initiatives	May substantially upgrade ITDC
Easier import of PRC engineers	Immediate	Expand knowledge base	
Attract foreign direct investment	Short term	Effect technology transfer	
Integrative science parks	Short term	Provide attractive work environment Cut cost of housing	
Launch technology fairs	Med. term cont. basis	Educate the public	
Themed exhibitions	Med. term cont. basis	Let company promote product Inspire youngsters	E.g. Office of the future,...
Encourage big co. to modernize	Med. term	Create opportunities for technology co.	Applies esp. to consumer services
Modernize government functions	Med. term	Create opportunities for technology co. Lead the private sector	Tech. developed can be applied elsewhere
Government-funded venture capital	Long term	Support new growing firms; Kick start venture capital industry in tech.	
A new stock trading system	Long term	Provide exit for venture capitalists Support growth of established firms	Separate small capitalization board or NASDAQ-type network

neers. The absence of engineers puts technology companies off even more. The situation grows continuously worse.

The immediate goal is to break the vicious circle and reestablish Hong Kong as a popular centre for the design of technology products. This will require that the government take aggressive steps to attract foreign direct investment, and that the government changes its immigration policy. A discussion of these steps is deferred to the next section of this chapter. The rest of this section addresses the more urgent need for increased technological expertise within the government.

At present, on technology matters, the government is advised by the Industry and Technology Development Council (ITDC). The ITDC is largely made up of government appointed representatives of the major manufacturing industries in Hong Kong. Within the ITDC, the committee that is primarily responsible for technology matters is the Technology Review Board (TRB). The TRB is made up of a small group of technologists from local and overseas universities, and of the upper management of local and multinational technology companies. The TRB does not meet often and all its members serve on an honorary basis. As an advisory body, the degree of the ITDC's influence on government policy is difficult for the public to gauge. When the ITDC was first established, the Financial Secretary of Hong Kong was the chairman. The public had high expectations for the ITDC at that time. Three years later, the chairmanship was passed to an executive councillor and the ITDC has since kept a low media profile.

If the government is to take a leadership role in developing technology, the technological knowledge pool within the government has to be greatly strengthened, and an advisory technology committee with high level access to the government has to be installed. The first condition would imply substantial strengthening of the Industry Department and the second condition would imply substantial upgrading of the ITDC.

The Industry Department could hire several teams of full time technologists, each of which would be held responsible for a technology area crucial to Hong Kong. Examples of such areas are integrated circuits, software, information and materials engineering. The department could then continuously conduct tasks that are presently being conducted on an ad hoc basis. It could periodically review the state of technology in Hong Kong industries, not only in the goods sector, but also in the service sector. As a result of such reviews, areas which enhance the co-ordination between different goods and service industries could be developed. The Industry Department could also promote Hong Kong to overseas technology companies and work out ways in which Hong Kong could best attract foreign direct investment in the targeted technology areas. This task is presently being carried out with limited resources and on a small scale. In conclusion, a much greater degree of knowledge support, manpower and recognition should be given to the technology sector of the government.

Once technological expertise within the government has been strengthened, the government's second task should be upgrading the ITDC. The role of a high level advisory committee in setting the direction of the Industry Department's technology policies is crucial. The advisory committee could review world trends and set out a strategy for the Industry Department to implement on a day-to-day basis. The role of the advisory committee is important because the "development front" of technology is very wide, consisting of some twenty to thirty distinct technology areas. Not all of these areas are relevant to the development of Hong Kong. Areas such as nuclear technology and aerospace technology are perhaps incompatible with the comparative advantages of Hong Kong. In sum, Hong Kong needs an advisory technology committee to target a few areas and steer the government's technology policies strategically in those areas.[1]

Foreign Direct Investment

Foreign direct investment is an important source of knowledge. When a foreign company sets up a design or manufacturing operation in Hong Kong, it brings with it machinery and experience operating this machinery. Foreign companies bring in production methods that are compatible with the cost conditions of Hong Kong. They also provide training for local engineers. Practical knowledge and training are important components of the knowledge pool that has been identified as a key prerequisite of a design-intensive strategy. These skills can only be acquired through on-the-job experience. They are not taught in universities and they cannot be found in books.

In Hong Kong, technology-related foreign direct investment has been weak in the past decade. Technology related foreign direct investments do not include two main types of foreign direct investment, namely, those which involve pure financial investments, such as investments in equity shares and in real estates, and those in establishing offices that provide marketing and sales support functions. Only technology-related foreign direct investments involve the useful transfer of technology. Some critics claim that no foreign technology companies have made a major investment in Hong Kong in the last five years. Political uncertainty has been put forward as a major reason behind this. This explanation is dubious, however, because many technology-related foreign direct investments have been made in China. The lack of interest among foreign investors can be better explained by economics than by politics.

The problem can be best analyzed by looking at investment from the foreign technology company's point of view. To the company, Hong Kong has a number of advantages: well developed telecommunications, good transport links with the rest of the world, low tax rates, free capital flow, no foreign exchange control and proximity to the China market. On the negative side, Hong Kong suffers from a number of deficiencies. First, it lacks experienced engineers. Second, it is costly for foreign companies to bring in engineers from outside, because housing prices are very high. Surveys show few employees in foreign-owned manufacturing companies are expatriates. In 1991, for example, 97 per cent of the employees in foreign-owned companies were local. Third, the working and living environment available is not very attractive. There are few industrial areas in Hong Kong that can boast a high quality living environment. Finally, the government has not been aggressive in striking deals to effect technology transfer. In Malaysia and Singapore, foreign direct investment is considered a top priority in state policy. In Hong Kong, the task is relegated to the Investment Promotion Division in the Industry Department.

To overcome Hong Kong's present deficiencies, fundamental changes in the attitude towards technology as well as in governmental policies are needed. In the present technology age, knowledge is the key infrastructure. Just as the government has supported the provision of infrastructure in the past, it should support the development of technology in future. This is not a matter of subsidy or intervention. Building up the knowledge base is an investment in the future. The government should play a leadership role in undertaking this investment.

To overcome Hong Kong's lack of engineers, the government should allow technology companies of good reputation to recruit engineers from China. China has attained such high standards in the fields of science and technology, and it has such a vast population of engineers who are eager to commercialize their skills, that it presents an excellent opportunity for Hong Kong. At the moment, because of the government's immigration policy, companies have to move some research operations to China. When it comes to its immigration policy, the government is particularly strict with Chinese workers. Under the current policy, companies can apply to bring in engineers from other countries much more freely than from China. The government fears that allowing companies to bring in engineers from China might lead to abuses. It is afraid that companies might bring in unqualified workers and effectively open a back door to Chinese immigrants. For this reason, what is suggested here is that only technology companies of good reputation be allowed the privilege of importing Chinese engineers. In practice, the government would have to exercise its judgement as to which companies qualify and which do not. This points again to the importance of building up technological expertise in the government.

Foreign companies hiring engineers from China may still be too costly if the cost of accommodation is taken into consideration. Moreover, the quality of the living environment in the industrial areas of Hong Kong is not particularly appealing at the moment. In this connection, the concept of a science park is interesting. The government has been studying the science park concept for a number of years. To date no definite work plans have been finalized. With hindsight, the earlier study of science parks was probably misdirected. The government studied science parks in terms of location choice and the intended areas of technology. The concept remained very much a cluster of buildings. It is hard to see how a few new buildings could by themselves draw foreign technology companies to Hong Kong. Looked at from a correct perspective, however, science parks could be the right way forward. Science parks should comprise both office and *residential* buildings. The residential units would be open only to employees of companies in the science park. Workers in the science park would not only find the arrangement convenient, but the concentration of workers in one neighbourhood could also generate close interaction among them and create an innovative atmosphere.

An attempt to attract foreign direct investment can only be successful if the government gives high priority to technology. It should realize that it is competing against other governments with much deeper technological backgrounds. The government has to be quick in identifying investor needs and in responding to those needs. To act fast the government may need to modify its hierarchical structure in order to reflect its dedication to technological development. As suggested earlier, strengthening the Industry Department and setting up a forward-looking, high-powered, advisory committee are moves in the right direction.

The discussion so far has addressed the short term goal of breaking the vicious circle of technological stagnation. The benefits to the economy at large of following the suggestions above may not be significant. It is quite likely that foreign companies form an enclave of their own. Everything in the enclave — the technology, the machinery, the materials, the manpower and so on — is imported. Products are exported to their own target markets and corporate earnings go to foreign investors. Does Hong Kong achieve any economic gains in the process? This question is troubling other industrializing countries as well. In the case of Singapore, even though the value of products produced in Singapore has increased rapidly, yet the increase has come about mainly as a result of the use of more machinery, not as a result of having learned how to use the machinery more effectively.

It is important to realize that the effort that goes into attracting foreign investment is basically an investment. The gains come when the knowledge diffuses throughout the economy, so that Hong Kong manufacturers begin to develop new products and designs. There will be real gains only when Hong Kong manufacturers can do indigenous research. The same thing applies to industrialized countries. Attaining a high technology level is only half the story. The more important second half is the way in which technology can be commercialized into desirable products. As was emphasized in Chapter 4, this process requires considerable entrepreneurial spirit and a competitive market. Herein lies the advantage of Hong Kong: the territory has long been a breeding ground for entrepreneurs and an adequate market of new products. Therefore, given its knowledge base, Hong Kong is in a better position to capture the potential of technology than are other industrializing countries.

In terms of government policy, the government can play a role in enhancing the technology utilization process and stimulating the growth of start-up firms. These are policies for the longer term. In the following sections, two broad strategic areas are discussed. They are the strategy for enhancing the public's awareness of technology and the strategy for improving the financing mechanism.

Awareness of Technology

At the moment, Hong Kong people's awareness of technology is generally rather limited. People are not inquisitive about the technology that drives the products they use. In the service sectors, in which frontier technologies in information engineering are being utilized, users seldom go one further step to find out how things work behind the scenes. In the school system, which is heavily examination oriented, little in the curriculum is designed to capture the imagination of the students and to make them see how they can use science and technology to enhance the quality of living. The best secondary schools students do not generally pursue scientific and engineering studies. Even among those who do, only a few aspire to work in the goods sector.

Despite Hong Kong's vibrant entrepreneurial spirit, the general lack of technological awareness is a real handicap to economic growth in the long term. The deficiency cuts both ways. On the one hand, only individuals capable of using technology creatively can create new products. On the other hand, in the product development process, entrepreneurs need the financial support of investors. To assess the potential of a technological venture, the investor needs a certain level of technological knowledge. Therefore, to make the emergence of new technology products possible, the awareness of technology must be raised across a broad front from users to investors.

Awareness of technology cannot be improved overnight. Improvement requires long term mass education. The government can take a leadership role in this respect. It can organize demonstration projects, technology fairs and exhibitions that introduce to the public the wonders of science and technology, how technology works in some common products, and how new technology can drive new products and serve society. In this vein, the governments of some countries run continuous exhibitions centring around themes such as "home of the future", "car of the future", or "office of the future". With some imagination and due attention to local needs, one may certainly add to the list "school of the future", "hospital of the future", "hotel of the future" and "restaurant of the future". Under each theme, the organizer might invite designers and manufacturers to put up exhibits and let the public see each product in action in a real living environment. The advantage of the arrangement is that it not only inspires the public and raises the level of its awareness of technology, but it also provides a venue in which technology firms can relate to the public and promote their reputation.

Another move worth considering is to have the government encourage large companies, such as airline companies, railways, electricity companies and so on, to upgrade their technology in running consumer services. As the real estate market, both private and public, catches a great deal of public attention, the government can encourage real estate developers to introduce more technology products into new housing. The government can take the lead in bringing in new technologies to modernize its own functions. Functions that are being carried out in government offices, such as fee collection, toll collection, import and export documentation, enquiries handling, information dissemination, energy conservation, material recycling, security systems, office automation, and a wide range of others can be modernized using new technology. Modernizing the government not only increases efficiency, but also sets an example for private firms to follow. These initiatives would not only enhance technology awareness, they would also create market opportunities for local technology companies. This is an important step towards fostering indigenous research.

Financing Mechanisms

In the longer term, when there is a presence of technology companies, a knowledge base of technology at the market level, and a high sensitivity to technological advances at the population level, then the financing mechanism of technology companies will become an important policy issue. Hong Kong has an advantage in this regard, because it has a high concentration of investment capital. The proper mechanism is needed, however, so that this capital is channelled to the companies that need it to grow.

The growth of a company can be divided into several stages each of which requires a different source of financing. In the first stage — the seed or start-up stage — the small business has only human capital, a few good ideas and maybe some business connections. To demonstrate the viability of its ideas, the business requires financing. If it is successful, it could come up with a prototype of demonstrable quality. The success rate of start-ups is usually rather low and therefore investments in seed companies are quite risky. More fortunate companies have a choice of selling out to larger companies or expanding the company to the second stage, the market stage.

In the market stage, the business takes the prototype and develops it into a marketable product. Manufacturing processes have to be worked out, materials have to be sourced, the market has to be studied and the product has to be promoted. These tasks require not only adequate financing but also good management. For some fortunate companies, sufficient earnings can be generated while the business grows from a seed company to a market company; outside financing is not necessary. But for many others, outside financing is required, and this is where venture capitalists play an important role. With funds from venture capitalists, a few companies may grow to really make a big impact on the market and have the ambition to diversify to other product lines or expand into foreign markets. A mature market company grows into a multi-product multinational enterprise. As it has a track record of high sales growth, it may be able to attract public investors and becomes a candidate for public listing on a stock exchange.

Generally speaking, as far as the source of financing is concerned, the younger the business is in terms of development stages, the closer investors and management must be. For start-up companies, investors are usually the management, their family members, or their close friends. The main reason for this is that the investor needs to know the educational and working background of the management, to be confident that it is honest, and to understand how it monitors the use of investment proceeds. The investor needs to know the management well, and preferably to have some means of controlling the management's opportunistic behaviour. As a result, those who invest in start-up companies are usually insiders.

The interesting issue is whether the government can justify using public money to invest in young businesses. In Hong Kong the government has already invested money in some start-up technology companies. Through the much publicized Applied Research and Development Grant, the government acts as a "business angel" and becomes the shareholder of these start-up companies. The amount the government has invested in each company has been low (usually not more than HK$1 million). According to some reports, response to the funding scheme has not been enthusiastic. The government was very cautious, as would be any investor, in approving applications. The approval process was long, involving committee decisions and outside reviewing. The quality of the applications, according to some observers, was generally low. Many applicants lacked awareness of developments in the world market and some applications were not regarded as viable in the Hong Kong context. The implication is that constraints on development have come more from a lack of technological knowledge than from a lack of financing. Some people also suspect that even good start-up firms may have second thoughts about applying to the government for support because of a potentially fatal information-leakage problem. These firms may perceive their future as highly dependent on a few good product ideas. To convince the government that the ideas are viable may necessitate giving away too much information.

It is probably too early to judge the success of the young Applied Research Grant scheme. It is evidently hard for the government to monitor the management of the assisted companies. Surely the government knows this, so most applications are rejected. Moreover, the depth of knowledge among businesses is not sufficient to create many viable products. So it may be concluded that the impact of such a public funding scheme is probably limited.

The financing of start-up companies is best left to private sources. Yet, the financing of companies in the middle stage of development may be another matter. These companies have a demonstrable prototype in hand. But the cost of bringing out a marketable product may be prohibitive. To adapt to the expansion in output volume, management needs to make adjustments in company routines and management style. In developed market economies, a wide range of venture capitalists are on hand to help such firms go through the transition and raise funds for the expansion. In the United States, there are an estimated 700 to 800 venture capitalists, many of whom specialize in technology companies. There is in fact a community of venture capitalists in Hong Kong, as well. Nevertheless, few of them are investing in local technology companies.

Most of them target Sino-foreign joint ventures and infrastructure development in China. The absence of technology venture capitalists in Hong Kong may simply be due to the perceived technological backwardness of Hong Kong. The problem may vanish when Hong Kong has reached a higher level of technological sophistication.

For the longer term, to facilitate the financing of growing technology companies, the government may consider setting up a venture capital fund. The venture capital fund would be publicly funded, but it should operate along commercial principles. The fund manager would run the fund as if it were a commercial operation. The fund would approve proposals on the merits of their quality of management and the viability and profitability of their product lines. Although such a fund would not target start-ups, it would help create an environment that fosters the emergence of start-ups by providing a milestone that they can work towards. For firms supported by the venture capital fund, the fund manager could provide more than just financing. He or she could provide market information; bring in foreign strategic alliances, suppliers and buyers; and help the young firm adapt its financial, personnel and marketing functions.

Some people may find the notion of a government owned, commercially run entity a contradiction in terms.[2] The real issue is whether the government can provide the right incentives to motivate fund managers. In the private sector, investors in venture capital are commonly institutions. They typically let fund managers decide in which ventures to invest. Fund managers take a 2 per cent to 3 per cent fee on the invested funds to cover overhead costs and earn a bonus if profits are made. If the fund goes bankrupt, investors lose their money and managers lose their jobs and their reputation. There is no immediate reason that the government cannot start a fund along these lines. It is risky business, but the risk that the government has to bear is not higher than that borne by commercial investors.

For established companies, the stock market has obvious advantages in channelling investor savings to finance expansion plans. The public can share in the growth of the company, though it takes it largely on faith that its investments are in good hands. In Hong Kong, as in other economies, only a small fraction of companies qualify to be listed on the Stock Exchange. Companies need to satisfy stringent conditions regarding their track record, asset size, information disclosure, validation of accounts and property ownership; they also need the support of underwriters in order to gain a listing status. These conditions safeguard the interests of small investors and minimize their chances of being cheated by unscrupulous directors of companies. The reduced risk does not come at no cost, however. There could well be promising companies that have genuine merits but that cannot gain access to stock market financing, because they are unable to meet the listing conditions. Technology companies, with their strong dependence on human capital rather than physical capital and their short profit history, are particularly handicapped in this regard.

In the United States, shares of many technology companies, including some of the largest ones, are traded in an over-the-counter market, not in a stock exchange. They are traded through the National Association of Securities Dealers' Automated Quotation (NASDAQ) system, a well established network of self regulated dealers. For companies, the NASDAQ's entry requirements are much less rigorous than are those of the main stock exchange, the New York Stock Exchange (NYSE). It is therefore not surprising that there are far more technology companies on the NASDAQ than there are on the NYSE. In fact, large technology companies such as Microsoft, Intel, Apple Computers, Sun Microsystems, AST Research, and Netscape could have qualified for the NYSE, but they have chosen to remain on the NASDAQ.

For the longer term, Hong Kong may consider establishing a share trading system with less stringent listing requirements than those of the HKSE. Such a trading system, whether it is a stock exchange for smaller capitalization companies or a network of dealers, would foster the growth of technology companies. It would also provide incentives for entrepreneurs to start new technology companies. Venture capitalists, who normally require an exit route to confine their investment horizons to a period ranging from three to five years, would also have a greater incentive to invest in technology companies.

For a technology company, obtaining a public listing status means more than just extra funding. Allowing company shares to be publicly traded, management gets a sense of how well the company is performing by looking at the stock price. When a company is large, outsiders frequently have a clearer view of its true worth than does upper management. Moreover, in the case of many technology companies, the best incentive scheme to motivate worker efforts could be an employee stock ownership plan. To make such a plan work a ready market for company stock is required, not only to provide a measure of corporate performance, but also to provide liquidity for workers. The public trading of shares also allows a company to restructure readily through mergers, acquisitions, spin-offs and divestitures.

Summing Up on Policies

The process of technological development is bound to be a long term one. Its ultimate goal is the establishment of an environment that nurtures the creation of design-intensive technology products. The initiatives have to come from entrepreneurs in the private sector. What is needed most is the revitalization of foreign direct investment. First, the government must build up its expertise in technology matters and consolidate the knowledge in the Industry Department. Second, it should strive to attract foreign direct investment in strategic technology areas. In this regard, the leadership role of a far-sighted advisory committee and the implementation role of the Industry Department are critical. To attract foreign investors, the government should allow them to recruit engineers from China and go ahead with the building of science parks which offer both office and residential space. For the longer term, nothing is more important than the diffusion of technology to the wider public. To enhance awareness of technology, exhibitions should be staged and many functions in the private and government sector should be modernized. To capture the financial strength of Hong

Kong, the government should consider in the longer term the enhancement of financing mechanisms to support technology companies. One initiative may be to set up venture capital funds to kick start a technology related venture capital industry. Another may be to introduce a share trading mechanism which allows easier entry for technology companies. Above all, the government has to give technology development high priority in the policy agenda and to show dedication to its long term implementation.

Before closing this chapter, it is necessary to address a few questions concerning the way in which government money should be best spent. These questions deal with whether the government should help displaced workers in the manufacturing sector, whether the government is already on the right track in developing technology with several recent schemes, and whether fiscal incentives should be granted to manufacturers to support research.

Employment Problem

Hong Kong still has some 400,000 workers in the manufacturing sector. Most people cannot see any prospect for these workers. Should the government do something to help them?

Many of theseworkers' skills do not command high wages. Manufacturers are unwilling to pay them wages high enough to keep pace with the increased cost of living. A large number of these workers move to the service sector. On the whole, government statistics show that unemployment has stayed low, implying that the capacity of the service sector to absorb these displaced workers seems rather high. We argue that developing technology and attracting foreign direct investment will not benefit these workers directly. The jobs created are mostly knowledge intensive and the displaced workers cannot be readily retrained to take them up. Even if the goods sector expands, manufacturers would probably do the manufacturing outside Hong Kong. The government cannot act against the changing economic reality. The benefits of technology development come mostly indirectly, through economic growth. Technology development would make possible the establishment of a viable goods sector, which would contribute to the long term growth of the economy. Such growth would in turn lead to an expansion of the service sector and the creation of more jobs.

Industrial Technology Centre, Industry Support Fund

The Applied Research Grant discussed earlier involves the government joining a start-up as an equity holder and sharing in the future successes and failures of the company. The Industry Support Fund is another source of funding, which aims at projects judged to be useful to the future development of Hong Kong manufacturing industries. Applicants to this fund may be manufacturers or university researchers. The nature of this scheme is very similar to that of the Research Grants Council, which provides funding on a competitive basis to university researchers. The main difference between the two is that the projects under the Industry Support Fund must have commercial market value, whereas those supported by the Research Grants Council are primarily basic research.

The Industrial Technology Centre is another recent government initiative intended to provide a nice working environment for start-up technology firms. It is governed by an independent board of directors and is intended to be financially self sufficient in the long term. The Centre provides leased office space, conference facilities, telecommunications and some consultancy services for companies. The Centre claims that its intention is to act as an incubator for start-up firms. When the firms grow and mature, they are expected to move out and start a bigger operation outside.

The impetus behind setting up the research grants and the technology centre is a good one, but its impact on technology development is rather small. The limiting factor is in the reliance on the existing pool of knowledge and manpower. To stimulate technology development, knowledge from foreign technology companies and manpower from China and elsewhere has to be brought into Hong Kong. Researchers benefit from interacting with each other and complementing each other. There has to be a critical mass of research manpower in order to launch Hong Kong onto a technology development path.

Fiscal Incentives

Some people argue that Hong Kong does not need to upgrade technology. If the government wants to support manufacturing and maintain a manufacturing sector in Hong Kong, all that is required are a few fiscal incentives such as tax credits for R&D expenditures and cheaper land granted to manufacturers.

Any plan for maintaining a viable goods sector in Hong Kong must take into consideration the development of technology. Better technological know-how, not lower costs, should be the primary incentive for manufacturers to carry out goods sector activities in Hong Kong. Giving manufacturers cheaper land does not by itself prevent them from moving activities out of Hong Kong. They could convert factory space into offices, showrooms and warehouses. Moreover, government survey data show that rental accounts for two to three per cent of total manufacturing cost. Therefore, even a large subsidy in land costs is not going to offset the higher local manpower costs. Cheaper land is unlikely to induce manufacturers to relocate manufacturing activities back to Hong Kong. What Hong Kong can gain from giving land subsidies to manufacturers is therefore difficult to see. The design-intensive strategy does not involve land subsidy in a central way. Design activities do not take up much land and therefore land cost is not a primary concern. Nevertheless, in the attempt to attract foreign investors, science parks that incorporate residential housing units to be leased to science park workers at concessionary rates should be built. The idea is the same as on campus housing being provided to university teachers.

Tax incentives on R&D expenditures deserve careful study. Even though the tax rate in Hong Kong is low, tax incentives do have a stimulating effect on R&D expenditure. However, there are a number of issues that need to be addressed if such tax incentives are to be given serious consideration. Some people may feel that such incentives would complicate the tax system. This problem aside, there are a few other complications. First, the stimulating effect of a tax incentive is stronger for large companies with high earnings than it is for small start-up companies with no profit record. The problem is not one of equity, but rather that small firms are usually more creative than large firms are. Giving large firms an advantage may discourage the formation of new start-up firms. Second, most manufacturing firms in Hong Kong do not carry out verifiable R&D work. What constitutes R&D under tax laws is certainly a problematic issue in the local setting. Third, many more technologically advanced manufacturers carry out R&D outside Hong Kong, for example in the mainland China, Taiwan and North America. Whether offshore R&D activities should fall under the incentive scheme is a difficult question. If they do, then manufacturers have no particular incentive to conduct R&D in Hong Kong and the diffusion aspect of technology development would suffer. If they do not, then the location in which manufacturers book R&D expenditures would have tax implications. The government has to incur extra costs to ensure compliance. The conclusion here is not that R&D tax incentives cannot stimulate R&D, but rather that the implementation of such a scheme requires in-depth analysis.

Notes

1. The choice of members in such a committee, whose recommendations could lead to substantial gains and losses for firms, is not a simple issue. As was pointed out by an anonymous referee, initiatives of the European Union in promoting generic technology areas (such as ESPRIT in information technology) are heavily influenced by big firms. Little money ultimately becomes available to small firms. If the government is to set up such a committee, it must choose members who can maintain a neutral, objective stance and who always put the interest of promoting Hong Kong's long term growth first.

2. There are such entities in Hong Kong — the Mass Transit Railway Corporation, the Kowloon Canton Railway Corporation and the Airport Authority. Such corporations are also quite common in Singapore.

CHAPTER 7

Conclusion

The Basic Law, the legal foundation that governs the way in which the Special Administrative Region government is going to administer Hong Kong, clearly sets out the importance of technological development. According to Article 118, the government is to encourage investment in new technologies and development of new products. If such investment and development is to take place, mere encouragement is not enough. The government needs to set its technology policy high on the policy agenda and to play a leadership role in spearheading various technology initiatives.

Technological development is closely related to the viability of the goods sector in Hong Kong. Keeping a strong goods sector in Hong Kong means maintaining an engine for export led growth. It also means strong support and quick response to the technological requirements of the service sector. Although a great deal of manufacturing activity has been relocated out of Hong Kong, mostly to China, Hong Kong still has advantages in various other goods sector activities, such as design and logistics. Hong Kong has the potential to become a design and nerve centre of global manufacturing.

Of particular importance is the design-intensive niche. Many modern products derive their value largely from their design. Design activities therefore capture a large part of the market value of a product. This is true for many hardware components in computers and electronic products; for all kinds of software; for many kinds of enhanced consumer products such as shoes, spectacle frames, clothing, watches, toys, electronic games and dictionaries; and possibly for many new consumer products that have yet to be created. In fact, some Hong Kong manufacturers have been successful in exploiting this design-intensive niche.

To pursue the design-intensive niche, Hong Kong needs to invest in technological development. Of top priority is the attraction of direct investment from foreign technology companies. This necessitates the creation of a favourable environment for foreign investors. The government has to play a major role in order to attract such foreign interest, and the public should be sympathetic in allowing the government to make large investments in order to reach its goal.

Looking at the situation from a different angle, it might appear that Hong Kong does not need to make all these investments in technological development. As China has a large pool of technological knowledge, an easier way forward may be to find ways in which this pool can be utilized to benefit Hong Kong. This argument is not altogether valid. While it is true that China is at the forefront in a number of technology related areas, much of its knowledge cannot be commercialized in its present state. Many engineers in China need market exposure and experience in actual production processes. Hong Kong manufacturers have a great deal of market exposure as well as insights into both Western and Chinese cultures. They also have the entrepreneurial spirit, the drive and the skills necessary to manage a business. The skills of Chinese engineers and Hong Kong manufacturers are therefore complementary. At the moment, some Hong Kong manufacturers are already using China's engineering knowledge to make their new product initiatives possible. In the future, Hong Kong should bring in more technology from foreign companies so that, with the help of China's basic research potential, more product ideas can come forth.

Direct investment from foreign technology companies give Hong Kong practical experience in managing the production of technology products. In the process Hong Kong will be kept up to date on new developments on the world scene. Yet, Hong Kong is not the only place that is going to attract foreign investment. Many industrializing countries are moving in the same direction. Does

Hong Kong have distinct advantages to offer? At the moment, it is hard to come up with any. In the past decade Hong Kong manufacturers have been preoccupied with transferring what they know into China, so the territory's technology level has remained stagnant. There have been few foreign technology start-ups in the territory in recent years. The government needs to devote considerable effort to turning the situation around; Table 6.1 summarizes our 11 suggestions.

Foremost on the government's agenda should be building up technological expertise within the Industry Department. A high level technology committee should be appointed to guide the Industry Department's technology related efforts. The immediate objective is to create an environment that attracts foreign technology companies to start manufacturing and design operations in Hong Kong. Foreign companies' concerns — the major ones being manpower, costs and living environment — will have to be addressed.

Regarding manpower, the local source is going to be limited. The government should allow foreign companies of good repute to recruit engineers from China. Regarding costs, the government should allow for the construction of residential units to accommodate engineers recruited from outside at below market level rental costs. Regarding living environment, the government should go ahead with the construction of the science park.

The efforts of the Industry Department and the attraction of foreign investment are best regarded as Hong Kong's investment in a technological future. The return on this investment comes when the technological knowledge acquired begins to diffuse to other parts of the economy. In order to create gains for Hong Kong, indigenous research is essential. The longer term goal is therefore to increase the level of the public's awareness of technology. Economic gains can be derived from the creation of new products that suit the local and Chinese markets. Here, Hong Kong has an advantage in that it possesses a significant local market and a vast market in China.

In the longer term, technological development may help Hong Kong develop as a financial centre. The presence of technology companies and increased awareness of technology would stimulate financing activities for these companies. These activities would cover not only technology companies in Hong Kong, but also those in China. When the conditions are right, venture capitalists, banks and other financial intermediaries would emerge to provide financial services. The government might also stimulate the mechanism by setting up its own venture capital fund. In time, the government may also set up a share trading mechanism which caters to small capitalization technology companies.

To conclude, seen from a broad perspective, Hong Kong already has unique advantages on the demand side of technology. It has a well educated population, a highly competitive market for most products and a high sensitivity to new products. Hong Kong people need to make use of all the available opportunities offered by technology to survive and prosper. The challenge is on the supply side. The market mechanism, which has worked well in other sectors, is not very effective in developing technology. The government should take a leadership role in technology development in order to sustain the growth of Hong Kong in the next century.

Bibliography

1. *Annual Digest of Statistics*, Hong Kong Government Printer. (Various years)

2. *Asian Venture Capitalist Yearbook 1994*, Asian Venture Capitalist Journal.

3. Gomory, Ralph (1992). "The Technology-Product Relationship: Early and Late Stages," in Nathan Rosenberg, Ralph Landau and David Mowery (eds.) *Technology and the Wealth of Nations.* Stanford, California: Stanford University Press.

4. Johnson, Chalmers (1982). *MITI and the Japanese Miracle.* Stanford, California: Stanford University Press.

5. Kao, Charles K. (1991). *A Choice Fulfilled.* Hong Kong: The Chinese University Press.

6. Kao, Charles K. and Young, Kenneth (1991). eds. *Technology Road Maps for Hong Kong — An In-Depth Study of Four Technology Areas.* Hong Kong: The Chinese University Press.

7. Kwong, Kai-Sun, Lawrence J. Lau, Tzong-Biau Lin (1996). "Productivity Changes and Deindustrialization in Hong Kong — An Empirical Analysis," mimeo.

8. Nelson, Richard and Sidney Winter (1982). *An Evolutionary Theory of Economic Change.* Cambridge, Mass: Harvard University Press.

9. Okimoto, Daniel I. (1989). *Between MITI and the Market.* Stanford, California: Stanford University Press.

10. Packard, David (1995). *The HP Way.* New York: Harper Collins.

11. Polanyi, Michael (1983). *The Tacit Dimension. Gloucester, Mass.:* Peter Smith..

12. Porter, Michael (1990). *The Competitive Advantage of Nations.* New York: The Free Press.

13. Romer, Paul (1993). "Two Strategies for Economic Development: Using Ideas and Producing Ideas." *Proceedings of the World Bank Annual Conference on Development Economics 1992.* Washington D.C.: World Bank.

14. Sexanian, Annalee (1994). *Regional Advantage — Culture and Competition in Silicon Valley and Route 128*. Cambridge, Mass.: Harvard University Press.

15. Simon, Denis F. (1995). ed. *The Emerging Technological Trajectory of the Pacific Rim*. New York: M.E. Sharpe Inc.

16. Young, Alwyn (1992). "A Tale of Two Cities: Factor Accumulation and Technical Change in Hong Kong and Singapore," *NBER Macroeconomic Annual* [Chicago].

Index

About the Author

Kai-Sun Kwong, PhD (British Columbia), is Associate Professor in the Department of Economics, The Chinese University of Hong Kong. He has served in a number of consultancy positions related to the industrial development in Hong Kong. In 1991 he served as consultant to the Industry Department of the Hong Kong Government on industrial pollution, in 1992 as consultant to the Technology Review Board of the Industry and Technology Development Council on technology assessment, and in 1994 as consultant to the World Bank on the mode of infrastructural development in Hong Kong. He has also been visiting scholar at the Department of Economics and the Stanford Institute of Theoretical Economics of Stanford University, California. He is the author of *Towards Open Skies and Uncongested Airports*, a publication of the Hong Kong Centre for Economic Research.

The Hong Kong Economic Policy Studies Series